WALTER C. PATTERSON

THE ENERGY ALTERNATIVE

Changing the Way the World Works

Boxtree

For Amory Lovins, Gerald Leach and Peter Chapman,
who started me thinking about energy.

First published 1990
© Walter C Patterson

Cover design by Vincent Design
Design by Sarah Hall
Typeset in Linotron Ehrhardt by Cambrian Typesetters
Printed and bound in Great Britain
by Richard Clay Limited, Bungay, Suffolk

for Boxtree Limited
36 Tavistock Street
London WC2E 7PB

British Library Cataloguing in Publication Data
Patterson, Walter C. (Walter Cram), 1936–
The energy alternative: changing the way the world works.
1. Energy. Supply & demand
I. Title
333.7912
ISBN 1–85283–284–3

Published in association with

CONTENTS

ACKNOWLEDGEMENTS

The Energy Alternative draws on the work and thought of analysts all over the world; even to attempt to list them would fill many pages. I must however record the gratitude I owe to many friends and colleagues whose ideas and actions have influenced my own. They include Amory Lovins, Gerald Leach, Peter Chapman, Czech Conroy, Mike Flood, Stewart Boyle, Simon Roberts, Mike Harper, Jose Goldemberg, Thomas Johansson, Amulya Reddy, Robert Williams, John Surrey, John Chesshire, Jim Skea, Steve Thomas, Jonathan Stern, Jane Carter, Mans Loennroth, Bent Sorensen, Andrew Warren, Ian Brown, Gerry Foley, Ariane van Buren, Robin Grove-White, David Olivier, Hugh Miall, Mark Barrett, Brian Wynne, Mick Kelly, George Henderson, David Merrick, Jim Harrison, Irene Smith, Malcolm Edwards, Chris Whatley, Dave Feickert, Brian Parkin, Andy Holmes, Lucy Plaskett, Chris Cragg, Gerard McCloskey, Frank Gray, Zeba Kalim, Paul Lyons, Max Wilkinson, David Lowry, Brian Price, Colin Hines, David Green, Steve Andrews, David Hall, Brian Brinkworth, Paul Hofseth, Erik van der Hoeven, Florentin Krause, Wilfrid Bach, Dominique Finon, Chen Zhi Hang, Lee Schipper, Jim Harding, Lester Brown, Chris Flavin, Cynthia Pollock Shea, Tom Cochran, Jacob Scherr and Denis Hayes; and when this book is irrevocably in print I shall doubtless recall with embarrassment many others I should have thanked.

   The Energy Alternative is a companion to the television series of the same title, commissioned by Channel 4 in Britain and co-produced by Independent Communications Associates and Grampian Television. As series adviser I have been informed, challenged and stimulated by my colleagues William Woollard, Ted Brocklebank, John Shepherd and Thelma Rumsey, whose energy seems boundless. My agent Abner Stein, and Sarah Mahaffy, Janita Clamp, Elaine Collins, Penelope Cream and Charles Phillips of Boxtree have seen *The Energy Alternative* into print with skill, dedication and impressive speed. My invaluable assistants Cindy Yates and Karen Brockett have wrought heroically to keep me from being submerged by my material. My beloved family, Cleone, Perdy and Tabby, are the best energy supply anyone could wish, and the best reason to use it.

Walt Patterson

Amersham, Bucks, UK
March 1990

PREFACE

Energy is dangerous. Using it can transform life on earth, for good or ill. We have long taken energy for granted; we use it without thought, and leave the thinking to others. This approach has served us well, many of us, for many years; but it now needs urgent reassessment, before the benefits of energy use are overtaken by its side-effects.

This book is an introductory guide to energy – what we do with it, how we do it, and how we might do it better. Chapter 1 sets the scene, outlining the issues that energy now raises. Chapter 2 describes what 'energy' is, and the role it plays in life on earth. Chapters 3 and 4 survey how energy use has evolved, and the controversies it precipitates. Chapter 5 shows how traditional energy planning is breaking down, and how an alternative is emerging. Chapter 6 describes the fundamental ideas underlying the alternative approach. Chapter 7 considers the opportunities for using energy more effectively, and Chapter 8 those for supplying fuels and electricity more acceptably. Chapter 9 contrasts energy issues in the industrial world with those in the Third World, where energy use lags ever farther behind. Chapter 10 contrasts the energy tradition with the energy alternative, its problems and its promise.

No single book can hope to do justice to a subject so vast. The particular examples cited herein are only a cursory sampling of activities now in train throughout the world; the appendices list sources that will provide many more examples and a wealth of further information.

One word of warning at the outset; you venture into this territory at your peril. Every human activity involves energy; and the innumerable issues that arise can take over your life. Two decades of experience leave this writer in no doubt: energy is habit-forming.

1
THE ENERGY DILEMMA

Strike a match.

You have just done something uniquely human. No other living creature, so far as we know, can start a fire at will. When our ancestors learned to control fire, they took the first step toward becoming the dominant form of life on earth. Controlling fire was the key to using what we now call 'energy'. If you can control fire, you can keep your body comfortably warm wherever you are, whatever the time of day, whatever the season. Instead of eating food raw you can cook it, vastly extending the edible menu. You can bake watertight pots and durable bricks. You can extract iron and other metals from the earth, and work them into strong and functional tools.

We have come a long way since the first intentional fire threw dancing shadows on the walls of a cave. We now know how to burn not only wood, dung and fat from animals and plants, but also coal, petroleum and natural gas. We can capture usable energy from falling water and blowing wind, and even release it from uranium. We can transform energy through incredibly intricate processes, at every scale from gigantic dam to silicon microchip. By using energy, we humans have transformed much of the natural world to suit ourselves.

We do not, however, know as much about using energy as we thought we did. With all our experience we still keep getting it wrong. An old adage is catching up with us: playing with fire can be dangerous. Indeed it may now be endangering our entire planet. We need to reexamine urgently this unique human ability, the ability to start a fire – that is, the ability to use energy on purpose. If we do not, it may soon be terminally out of control.

Think about your burning match. What could you do with it? You could light a bonfire to bake your potatoes; or you could raze a forest. You could ignite your kitchen gas-ring; or you could level a shopping centre. You could fire up the coal in a factory boiler; or you could set off a disastrous explosion in a mine. Hazards like these are essentially immediate and local. People have lived with them for many decades. We try to minimize them; but we accept them in exchange for the obvious benefits we gain by using energy. Within our lifetimes, however, we have become acutely aware of other hazards, more subtle and indirect, more long-term and far-reaching.

Anyone who reads the news headlines will long since be familiar with the dismaying litany. Choking smog turns urban air a dirty orange; eyes water and lungs inflame in Los Angeles, Athens, Madrid, Mexico City, indeed everywhere that too many cars burn too much petrol under too much sunlight. 'Acid rain' laden with sulphur and nitrogen compounds from burning coal and oil poisons rivers, lakes and soil; fish die and woodlands wither. The explosion of the Chernobyl nuclear plant in April 1986 showed that even a single accident can affect geographical areas thousands of kilometres away, for years to come.

In the 1990s, however, all these unintended consequences of energy use, alarming though they be, have been overtaken by yet another – the most awesome yet. Scientific evidence indicates that we are now upsetting the climate of our planet. Six of the hottest years ever recorded occurred in the 1980s, bringing droughts, violent storms, floods and extremes of weather to many parts of the earth. Leading scientists concur that we may be seeing the first warning symptoms of a so-called 'greenhouse effect'. Gases we have released into the atmosphere are behaving like a greenhouse surrounding the planet. When the sun shines into this greenhouse it raises the temperature inside, producing unpredictable changes in complex planetary systems like the atmosphere and oceans. Our use of energy is a key cause of the disturbance; one of the main 'greenhouse gases' is carbon dioxide, released by burning coal, oil or natural gas.

Every time you strike a match, turn on a light or press the starter on your car you complicate matters that little bit further. Your impact alone is negligible – but you are one of five billion people on earth. Many millions of your fellows are also turning on lights, pressing starters, and using energy in an endless variety of other activities, in households, vehicles, offices, factories and wherever else they happen to be. To be sure, many of the five billion people on earth have little opportunity to use energy the way we do in industrial countries; but they can claim with justice that they are entitled to share the consequent benefits we now take for granted. If all five billion of us, and the additional billions yet to come, try to use energy the way we do in industrial countries at the onset of the 1990s, the outcome may be catastophe.

Today, on average, each person on earth uses about two-thirds of a tonne of coal, three-fifths of a tonne of oil, 350 cubic metres of natural gas and 2000 units of nuclear and hydroelectricity in a year. That 'average', however, is wildly misleading. The average person in the industrial world uses about ten times as much as the average person in the so-called 'Third World'. Even that average is misleading, because it lumps together those who live in cities and those who live in rural areas;

and so on. Indeed, even this sort of description, in terms of oil, coal, gas and electricity, is misleading, as we shall see. Nobody anywhere actually uses an 'average' amount of energy. Each user is different and distinctive – you included.

Most of us take energy pretty much for granted; but many millions of people make energy their business. Some supply fuels or electricity; some supply the hardware to use the fuels or electricity; some design new hardware to use or supply energy. Some analyse resources, prices and patterns of energy use and supply, to anticipate and plan for future developments. Whatever their particular involvement in the energy business, they have had to devise manageable ways to think about it. Never before have they had to think as hard as they do now. The conclusions they reach and the decisions they take will affect every aspect of your life, from the bills you pay to the air you breathe.

As the world enters the 1990s, the global energy scene is a tumult of confusion and controversy. Key assumptions and ground-rules no longer hold; but convincing alternatives have yet to emerge. Only fifteen years ago, in the mid-1970s, most commentators were worried that oil, gas and coal would become ever scarcer and more expensive – that supplies of these fuels could not rise rapidly enough to fulfil the demand for them. In late 1973, within only a few weeks, the Organization of Petroleum Exporting Countries (OPEC) quadrupled the international price of crude oil, triggering panic throughout the industrial world. Governments initiated grandiose programmes to find substitutes for oil, anticipating that its price would continue to climb; industry and commerce succumbed to a deep recession. In 1979 the overthrow of the Shah of Iran, followed by a grim and debilitating war between Iran and Iraq, precipitated a second panic over oil supplies from the Middle East; gun-toting motorists in the US even fought pitched battles around the petrol pumps in service-station forecourts.

Only a decade later, however, oil, natural gas and coal are more abundant than ever. Their prices are not too high but too low – gratifying for industry and commerce in the short term, but too low for suppliers, too low to remain stable and reliable for suppliers and users alike, and too low for the global environment. In the 1970s, high fuel prices encouraged fuel-users to change their more wasteful and extravagant habits, in order to reduce costs. In the 1990s, low fuel prices have weakened this resolve. Fuel-use has once again begun to increase rapidly – aggravating problems like acid rain, toxic waste disposal and the greenhouse effect.

Faced by this intensifying dilemma, energy planners in governments, companies and international organizations like the United Nations and the European Community are striving to recast their plans and policies, to cope with the deepening uncertainties they now confront. Their task is unenviable. Governments must find a way to establish ground-rules –

taxes, environmental regulations, planning laws – that will reconcile the need to protect air, water and land with the need to keep the lights from going out. Companies must try to anticipate the fuel and electricity supplies and energy hardware that will be required in the decades to come, and how much they will cost, in order to channel their investments in the best directions; no company wants to find that it has staked its future on buggy-whips when the horseless carriage is about to appear.

International organizations must mediate between different nations, whose interests often conflict. If one national government imposes more stringent environmental controls than another, it may find that its manufactured goods become more expensive than those of less environmentally scrupulous competitors. On the other hand, if international environmental standards are lax enough to suit the least scrupulous country, the quality of air, water and land will suffer – and so will those who breathe, drink and live on them. Pollution pays no attention to political borders; why should one nation clean up its act if its neighbours do not? The atmosphere and the oceans are 'commons', accessible to everyone; but all too often a country seeks to gain an unfair economic advantage by offloading the damaging side-effects of its energy use.

Accordingly, the energy policies that governments, companies and international organizations adopt and implement in the 1990s will have a fundamental and immediate effect on their citizens and their customers. Government, corporate and international policies will determine how much you will pay for fuels and electricity. They will influence the kind of homes, cars and consumer goods you can purchase, and how much they will cost to buy and to operate. Above all, the policies will affect the health and vitality of the environment in which you live – both local and global. Even in your own interest you owe it to yourself to keep an informed eye on the policymakers – and to let them know what you think.

They are now more sensitive to public opinion than they have ever been. They know that energy breeds conflict and controversy, at every level. OPEC member countries clash over oil-production quotas; coal companies clash with their miners; gas suppliers clash with disgruntled clients; local objectors clash with power-plant constructors; governments clash over pollution controls; the entire energy panorama is laid out in battlefields.

Nor does the recent record of official energy plans and planners inspire much confidence. In the past two decades, both national and international 'forecasts' of future fuel and electricity use have proved in the event relentlessly and invariably wide of the mark. In the past fifteen years, the old guidelines for energy forecasting have been comprehensively discredited, as we shall describe in Chapter 5. As yet, however, energy analysts have not come up with fully convincing replacements. The field is in ferment. Computers churn out roomfuls of printouts; scientists, engineers, economists and accountants wrestle with incomplete, inconsistent and often contradictory findings about where the human use of energy may be going. Energy planners cannot make up their minds about what they can assume, or what they must consider. If no one anywhere really is the mythical 'average energy user', how can you anticipate what energy users will do, and prepare accordingly – without wasting enormous sums of money, or devastating the environment, or just making a fool of yourself?

If you are an energy planner or policymaker, one point is clear: it is impossible to do everything. You can contemplate today an extraordinary shopping list of options – different technologies for using and supplying energy, different financial arrangements, laws and regulations, different organizations and institutions. But you must choose among these options, and hope you are making the right choice. The same challenge arises at every level, from domestic to international. Do you select this deep-freeze instead of other options, even though it costs more? Do you select this pollution-control standard instead of other options, even though it costs more?

The choices are further complicated because no option is guaranteed trouble-free. In the mid-1980s you could have selected your new deep-freeze, happy in the knowledge that its operating system contained only the safe, inert chemicals called 'chlorofluorocarbons' or CFCs. Now we know that CFCs damage the protective ozone layer that shields the earth from the sun's ultraviolet radiation. Even the most apparently innocuous choice we make may subsequently backfire on us – scarcely a reassuring realization amid so much uncertainty.

Yet never before have we been offered such a sumptuous shopping list of energy options. We can choose among different fuels and their suppliers, different supply technologies and their suppliers, different energy applications, their hardware and their suppliers. All the options have advantages and disadvantages, some evident and some much less so. We have both too much information and not enough. Advertising, official reports from governments and international organizations, unofficial reports from universities and environmental organizations, analyses from investment brokers, articles in print and broadcasts on TV and radio – keeping up with the outpourings is more than a full-time job; but major gaps remain unfilled.

As yet we simply do not know enough about the comparative costs and performance of different energy technologies, or about their environmental impacts, or about the feasibility of competing options. Enthusiastic salesmanship, whether for nuclear power stations or for windmills, is no substitute for actual experience, as we have all too often realized too late. Despite the uncertainties, however, choices must be made. Even to sustain our present patterns of energy use, policies and plans must be drafted and implemented, hardware designed, ordered, installed and operated.

Within the daunting range of possibilities, two approaches can be discerned. We can characterize them broadly as 'the energy tradition' and 'the energy alternative'. The energy tradition tends to focus on supplying fuels and electricity; the energy alternative tends to focus on using them – how and for what purpose. The energy tradition tends to deal in averages and aggregates; the energy alternative tends to deal in specifics and particulars. The energy tradition evolved gradually, over more than a century; the energy alternative has emerged within the past two decades. In subsequent chapers we shall track the two approaches, and consider their diverging implications.

To reach the best available decisions we need to understand clearly what we know and what we don't know. That must encompass everything relevant: basic engineering of energy hardware to get the best performance from entire systems; new materials, new processes and new combinations; their impact on society, and on the local and the global environment; and how different packages of choices may evolve in years to come. As we pick our way among the energy options and uncertainties – scientific and technical, economic and financial, social and political – one key question becomes paramount. Who is to make the energy choices that will determine the sort of world we pass on to our grandchildren? Who decides, and how?

Think about it the next time you strike a match.

2
WHAT IS 'ENERGY'?

'Energy is eternal delight.' So wrote the English poet and mystic William Blake in 1790. Two hundred years later, however, the ecstatic purity of Blake's dictum has been more than somewhat muddied. To Blake, 'energy' was a manifestation of the vitality of the world – an essentially mystical concept. He would have recoiled from any thought of 'defining' energy, much less measuring it and describing it with numbers. But the 'dark satanic mills', the coal-burning factories that Blake deplored, were going to change 'energy' from 'eternal delight' to something much more problematical.

According to the Oxford English Dictionary, 'energy', the word itself, made its first recorded appearance in 1599, meaning 'force or vigour of expression' – a sense undoubtedly congenial to Blake. It is equivalent to the Greek word 'energeia', used by Aristotle, and comes from the Greek work 'ergon' meaning 'work'. Other meanings gradually accrued. By 1665 it meant 'power actively and efficiently exerted', and by 1667 'ability or capacity to produce an effect'. The word 'energy' described an attribute recognizable by its consequences. You could tell that a person had 'energy' by watching the person in action. A person with energy could move swiftly, or lift heavy weights – and might get hot and sweaty doing so. You could even say that one person had 'more energy' than another; but the comparison was vague and imprecise. We still use the word 'energy' like this in everyday speech; it conveys a meaning that is adequately clear and intelligible for everyday use. But the association of a person's 'energy' with movement, force, work and heat is a clue to the much more potent and precise application of the concept of energy, which emerged less than 150 years ago.

DISCOVERING ENERGY

By the seventeenth century, even as the word 'energy' was coming into parlance, thinkers using a new approach called 'natural philosophy' – what we now know as 'science' – were seeking new and more powerful ways to understand and describe the world. One key idea was comparison not only with words but with numbers. Instead of simply calling one horse 'short' and another 'tall', you counted the number of your hands'-breadths from the ground to its shoulder: one horse might be 14 hands high, another 15, another 16. Provided you always used the same hand as 'one hand' – a 'unit' of horse-height – you could compare as many horses as you wished, and rank them in precise order of height. The same technique could be used for comparing lengths of cloth, or sacks of grain, or even intervals of time, if you picked a suitable unit; each inversion of an hourglass, for instance, would indicate one more hour passed. This idea of comparison by counting units was called 'measurement'. Initially, it was a purely practical procedure, enabling traders and their customers to agree how much of a commodity was being supplied and purchased, on the basis of 'weights and measures' mutually acceptable – an 'ell' of cloth, a 'pound' of turnips, an 'acre' of land. However, in due course measurement also fostered a much more precise description of natural circumstances and events.

Scientist like Galileo Galilei and Isaac Newton set about developing ways to compare phenomena like movement not only with words like 'slow' and 'fast' but also with numbers: '10 feet per second', '100 feet per second', and so on – and to represent these relationships in the symbolic language called 'mathematics'. In the following century Daniel Gabriel Fahrenheit and Anders Celsius devised ways to compare the subjective sensations of 'hot' and 'cold' with numbers, by means of 'thermometers' calibrated to give a numerical scale of 'temperature' in steps of 'degrees'. Antoine Lavoisier and Pierre Laplace invented 'calorimeters' that could measure the amount and flow of heat between hot and cold bodies, in 'calories', from the Latin word 'calor' meaning heat.

While the scientists were working out consistent and coherent ways to describe the workings of nature, practical engineers were applying their own common-sense understanding. They had known for millenia that the efforts of human and animal muscles could be enhanced by using devices like levers, sloping ramps or 'inclined planes', ropes and pulleys and other so called 'machines' to make 'work' easier – lifting heavy objects and moving them from place to place. Gradually the concept of 'work' emerged as another quantity that could be measured and described with units and numbers. Lifting a measured weight a measured height required so much 'work'; lifting twice the weight the same height required twice as much work; lifting the original weight twice as high also required twice as much work.

Moreover, you could do the work by using not only animate muscles but also, for instance, a falling weight, or water running over a mill-wheel, or other inanimate means. By the beginning of the nineteenth century scientists had recognized that muscles, falling weights, running water and other natural systems contained what you could think of as 'stored work', that you could use to apply forces and make things move. In 1807 the British physicist Thomas Young proposed that this stored work be called 'energy'. It was to prove a concept even more potent than Young could have imagined.

One unexpected consequence arose from the activities of Benjamin Thompson, the brilliant scientist-administrator who founded Britain's Royal Institution. In the course of his multifarious career Thompson spent a decade in the service of the Elector of Bavaria, who granted Thompson the title Count von Rumford. As head of the Bavarian War Department Thompson supervised the manufacture of gun-barrels; and he was struck by the fact that drilling out the holes in the metal castings made the metal hot. He surmised that the work done to turn the drill was related to the heat acquired by the casting; but precise measurements were too difficult to be accurate. Not until 1843, more than fifty years later, was Thompson's conjecture proved beyond doubt. In a series of elegant and scrupulously careful experiments, the British physicist James Prescott Joule demonstrated that the amount of work done by a falling weight turning paddles in a water-filled vessel was exactly proportional to the amount of heat transferred to the water.

Joule's experiments showed conclusively that work and heat were both manifestations of the same physical property: each was 'energy', and one manifestation could be transformed into the other. Joule unified what had been to that time widely disparate aspects of nature, and revealed the real potency of the concept of energy. As scientists explored its implications, they came to acknowledge energy as a fundamental attribute of the universe, a unifying principle allowing us to see the connections between the most apparently different circumstances and events.

Countless experiments, of every variety, in due course convinced scientists of a remarkable fact: energy is never created, and never destroyed. Indeed this is probably the best-known popular fact about energy – the 'law of conservation of energy'. Contrary to popular belief, the total energy consumption of the world is ZERO. No matter what official statistics may tell you, no one – no householder, no driver, no factory worker – 'consumes' energy. Energy is 'converted', from one form to another; but the quantity of energy always remains the same.

Energy, as we now know, manifests itself in a profusion of ways. Suppose, for instance, you shoot an arrow at a target. When you bend the bow, 'chemical energy' from the food you have eaten is converted into 'mechanical energy' of your contracting muscles. It is then stored briefly as 'elastic energy' in the bent bow and the stretched bowstring. When you release the bowstring, the stored elastic energy is converted into energy of motion – 'kinetic energy' – of the arrow. When the arrow comes to rest in the target, the arrow's kinetic energy is converted into energy of deformation of the target, and by friction into heat energy; the arrow and target get slightly warmer. Some of the energy will also be converted into 'acoustic energy' – the whistle of the arrow through the air and the 'thunk' as it hits the target; and so on. Throughout the whole process the total quantity of energy involved remains unaltered. This is precisely what makes energy such a useful scientific concept. By keeping track of the total quantity of energy throughout a conversion process, we can establish measurable connections between physical phenomena whose relationships seem otherwise remote or complex, like the bend in the bow and the distance the arrow travels.

We can measure the quantity of energy involved by watching the effect produced when it is converted: how much a bow can be bent, how far an arrow can be shot, and so on. However, as we have seen, until the 1840s, few people realized that mechanical motion and heat were both manifestations of the same physical concept, or that one could be converted consistently into the other. People therefore came up with an assortment of 'units' for measuring energy, depending on which manifestation was being measured: 'foot-pounds', 'calories', 'British thermal units' and many, many others. The jumble of historical energy units is a grotesquely clumsy vocabulary to inflict on a concept that ought to foster simplicity and clarity. All the various energy units are convertible one into another, but the arithmetic entailed is infuriating; even energy experts get slightly shrill when discussing energy units.

We are, however, stuck with them, as we are stuck with many other awkward idioms in the language of energy; see Appendix A. The energy units of the international scientific system, the Système Internationale (SI) are gradually – very gradually – becoming the standard in most parts of the world. Fittingly enough, the basic SI unit of energy is named after the scientist who showed its universal applicability: it is called the 'joule', pronounced – some might say appropriately, considering the value of the concept – 'jewel'. To lift a bag of granulated sugar, a mass of one kilogram, its own height, about a tenth of a metre against the pull of gravity, you must convert one joule of energy. To lift the bag 1 metre you must convert 10 joules; and so on. If you are a fifty-kilogram woman going up a flight of stairs three metres high, you must convert $50 \times 3 \times 10$ $= 1500$ joules of energy just to lift yourself from the bottom of the stairs to the top. As you can see, the joule is a small unit. When discussing the quantities of energy being converted by natural systems and human activities around the world, in order to have conveniently manageable numbers we shall talk in multiples of the basic unit, identified by their prefixes. Just as 1000 metres equals 1 kilometre, so 1000 joules equals one kilojoule; and so on, like this:

        1000 joules = 1 kilojoule (KJ)
        1000 kilojoules = 1 megajoule (MJ)
        1000 megajoules = 1 gigajoule (GJ)

1000 gigajoules = 1 terajoule (TJ)
1000 terajoules = 1 petajoule (PJ)
1000 petajoules = 1 exajoule (EJ)

Thus, the woman climbing the stairs converts 1500 joules, or 1.5 kilojoules.

The rate at which energy is converted is called the 'power'. Converting one joule of energy per second delivers a power of one 'watt'; the unit of power is named after James Watt, the British scientist and engineer who provided the key to a crucially important form of power, as we shall see. Just as the joule is a small unit of energy, so the watt is a small unit of power. We shall find ourselves talking about thousands, millions and billions of watts – that is, kilowatts, megawatts and gigawatts. Even when you are asleep, your body is converting energy – breathing, pumping blood, sending nerve-impulses – at a rate of about 100 watts, comparable to a bright light-bulb. If the woman mentioned above climbed the stairs in three seconds she would be converting 1500 joules in 3 seconds: that is 1500/3 = 500 joules per second, or 500 watts.

CONVERTING ENERGY

Every change – EVERY change, of anything, anywhere, any time – is accompanied by conversion of energy; but the total quantity of energy involved never changes. What changes is the net QUALITY of the energy. The quality of energy depends on its organization and its concentration. The more highly organized or more concentrated the energy, the higher its quality. You can give extra energy to an arrow by firing it from a bow – organized energy, in which all the particles of the arrow move the same way; or by heating it over a flame – disorganized energy, in which the particles of the arrow move back and forth in random directions. We now know that what we have long called 'heat' is actually disorganized energy. Its concentration is called 'temperature': the higher the concentration of disorganized energy, the higher the temperature – for instance, the hotter the arrow.

As you might expect, in any energy conversion process, although the total QUANTITY of energy remains unaltered, the net QUALITY of the energy decreases. The kinetic energy of the flying arrow is high-quality energy because it is orderly and well-organized: all the parts of the arrow move in a coherent fashion together. When the arrow imbeds itself in the target, this organized energy is converted and scattered in all directions, as the arrow bends and tears the target and rubs against it, making arrow and target slightly warmer. Even this small amount of locally concentrated heat energy then spreads out through the material until everything is once again at the same temperature as the surroundings; the original high-quality energy is now disorganized and diluted to such an extent that it has become imperceptible. Any energy conversion process has similar consequences. Energy that is highly organized becomes less so; concentrated heat energy at a high temperature spreads out and becomes cooler.

Fortunately for us, however, we have a treasure-house of high-quality energy available. Energy makes the world go round; and almost all the energy that does so on earth comes from the sun. The heart of the sun pours out high-quality energy in the form of sunlight, visible and invisible, which streams into space in all directions. A minute fraction of this sunlight is intercepted, 150 million kilometres away, by the earth. Some of the sunlight that strikes the outer atmosphere of the earth is at once reflected back into space. Some is absorbed in the atmosphere; some reaches the surface of the earth, directly and indirectly. The high-quality sunlight energy is converted by many different processes in the atmosphere and on the surface of the planet. Some sunlight energy is converted into heat, warming the air, the water and the soil. Some drives elegant chemical reactions in green leaves, which store the sunlight energy as chemical energy in energy-rich molecules like starch and sugar. Some sunlight energy evaporates water from rivers, lakes and oceans. As the water vapour moves through the atmosphere, sooner or later it condenses again into rain or snow, giving up the energy acquired during evaporation; you may have noticed that the air feels warmer after a rainfall. Currents in the oceans and winds in the atmosphere carry warmer water and air from place to place. Thus the sunlight energy that reaches the earth during any given hour is soon apportioned all over the planet, as 'ambient' energy – that is, the energy of the local environment. This ambient energy in your surroundings is converted continually from one form to another, its net quality decreasing as it drives the weather, the growth and activities of plants and animals, and all the other physical, chemical and biological processes taking place everywhere.

However, the earth does not acquire this energy to keep. Energy tends to move from a warm place, where the heat energy is concentrated, to a cooler place, where the heat energy is more dilute. In outer space the heat energy is very dilute indeed; the temperature goes down to about 270 degrees below zero Celsius – only 3 degrees above absolute zero. Needless to say the earth is considerably warmer than this; so the earth, too, sends a stream of energy – 'earthlight' – in all directions into space. The apparent temperature of the earth, when viewed from space, is determined by the balance between the high-quality sunlight arriving on the earth and the low-quality 'earthlight' – invisible infrared light – leaving the earth. The earth's 'average' temperature is such that the energy arriving is offset by the energy leaving, so that the 'average' temperature remains the same. If energy were accumulating on the earth, its 'average' temperature would rise until the amount of energy leaving was again in balance with the amount arriving. Of course the earth's 'average' temperature when viewed from space is an average over a wide range of prevailing temperatures in various parts of the earth's atmosphere and on its surface – from more than 50 degrees below zero Celsius to more than 50 degrees above. The whole system is extraordinarily complex, and as yet far from well understood. But its behaviour has far-reaching implications for our future attitude to energy, as we shall see.

More than 99 per cent of the energy conversion that takes place on earth involves sunlight energy in one form or another. The quantity of this energy arriving and leaving is enormous; and almost all the conversion it undergoes takes place willy-nilly, with no intervention by human beings. The motions of the atmosphere and the oceans, the progression of the seasons, the variations of outdoor temperatures from place to place, from hour to hour and from year to year, the life and death of wild plants and animals, all are driven by converting the ambient energy of sunlight.

We humans, too, rely on sunlight for almost all the energy conversion processes of importance to us. Sunlight keeps the average temperature of the earth almost high enough for human bodies to survive and function as living organisms with no additional assistance; look at any popular beach resort in summer. Sunlight energy stored in plants provides food for people, either directly when we eat the plants or indirectly when we eat animals that have eaten the plants. People, including energy experts, tend to call energy 'useful' only when they notice that they are using it; but none of us would survive without the continuing support of the natural ambient-energy conversion processes we take for granted. The energy we 'use' and call 'useful' – because we intervene consciously in its conversion – is a modest, not to say trifling addition to the natural ambient-energy conversion processes on which we unconsciously rely.

PORTABLE ENERGY

Nevertheless, the human use of energy differs in one unique and telling respect from the use of energy by all other living things. Human beings have learned to use 'fuel'. The word 'fuel' comes from old French 'fowaille', which comes in turn from low Latin 'focale' and Latin 'focus' – meaning 'fireplace'. A fuel is 'material for a fireplace': material whose energy-content can be mobilized under human control, and converted at places, times and rates chosen by humans. However, our unique ability as humans lies not merely in recognizing the concept of fuel as a store of accessible energy. The significance of the myth of Prometheus, the human acquisition of fire, was not the mere discovery of the process of burning; fire occurs in nature, as a result of lightning, spontaneous combustion and other causes. The significance of the myth was the discovery that a human could CONTROL fire: that is, a human could start a fire at a chosen place and time. No other living creature has mastered this skill. Other living creatures grow and shed fur or other body coverings, to retard or accelerate the loss of heat from their bodies, and seek or build shelters to provide comparatively stable temperatures around them. But no other creature can mobilize at will energy stored OUTSIDE its body, to initiate a controlled process of energy conversion, and raise the local temperature to a level much higher than the ambient temperature. Of all the characteristics that distinguish humans from other living creatures, the ability to use fuel may be paramount – possibly even more distinctive than our development of language. Not even the great apes use fuel.

No one will ever know just how humans learned to use fuel. Perhaps, one day, one of our ancestors found a tree burning after a lightning strike, and liked the warmth and light it provided. Our ancestor might have noticed that the fire could spread from one branch to another; perhaps, as the fire began to die, our ancestor took the initiative and placed another dry branch nearby, to see if it too would begin to give off warmth and light. It seems likely that for a long time humans could only feed and cherish fires started by natural occurrences; actually starting a fire must have been accomplished much later. Even today you will have great difficulty starting a fire using only natural materials, if you do not resort to a sophisticated human device like a match or a magnifying-glass. If you try, you will find that you need dry fuel, preferably finely divided and of low density – for instance dry leaves. You must heat this fuel until it is much hotter than ambient temperature, when the fuel will begin to react rapidly with the oxygen of the air. This reaction, once started, gives off more heat, until the reaction is occurring throughout a significant volume of the fuel – it has 'ignited' and started to burn. Your problem is to bring about the initial temperature rise, to start the reaction.

If you have only natural materials at ambient temperature, you have two possible ways to raise the temperature of a material until it ignites. You can take two suitable rocks and bang them together, hoping that the mechanical energy of your muscles will be concentrated into an abrupt and violent tiny fracture at the point of contact of the rocks, producing a speck of very hot broken rock called a spark. If this spark lands in the finely-divided fuel – say the dry leaves – the spark may transfer its heat to the fuel and raise its temperature to ignition. Of course you may have to spend a long time banging the rocks together before you get a hot enough spark to land just right. Alternatively, you can take two pieces of wood, suitably shaped, and rub them vigorously together so that the mechanical energy from your muscular motion will be converted into frictional heat at the point of contact between the pieces of wood. If you place your finely-divided fuel at this point you may be able to raise its temperature enough to ignite it. Of course your arms may get very tired in the process. In either case the concentrated high-temperature heat you have laboriously generated by your efforts may leak away into the surrounding air too fast, so that the fuel never gets hot enough to ignite. In retrospect, the human achievement of learning to start a fire is very impressive indeed, as you can confirm for yourself the next time you take a walk in the woods.

Nevertheless our ancestors did learn to start a fire. In doing so they acquired an ability whose consequences have been incalculable. The primary result of this ability is that a human can create, at will, a local temperature far higher than that found normally in the parts of nature accessible to humans. Access to warmth and light independent of sunlight has allowed people to survive even in polar winters, and colonize almost the entire planet. Furthermore, such high temperatures foster many interesting phenomena. We can cook food that would be inedible or unhealthy uncooked. We can harden and glaze earthenware vessels to make them watertight and heat-resistant. We can smelt ores to yield metals, and shape the metals into tools. Worked metal tools in turn make it easier to prepare fuels and ignite them. In short, the ability to start a fire made it thereafter progressively easier to start a fire.

The ability to use fuel also allowed people to stockpile a store of energy that could be tapped as desired. They could carry this stored energy from place to place and use it when and where they wished. Fuel was thus not only an energy store but also an 'energy carrier'. Throughout the prehistoric eons humans discovered a wide variety of materials that would burn – wood and other dried plant materials, dried animal dung, animal and vegetable fats and oils, peat and coal. They also built increasingly elaborate structures and devices to utilize the heat from fuel: ovens, kilns, furnaces, braziers and so on. With tools made from metals recovered and shaped by using fuels, people also created structures and devices that could control for human purposes the conversion of natural ambient energy carried by falling water and blowing wind: water mills and pumps, windmills and sails.

ORGANIZING ENERGY

However, despite the rapidly expanding opportunities for using fuel, one major limitation remained. Energy converted by burning fuel manifests itself as heat. Heat, no matter how hot, is disorganized energy. The energy of material motion, for instance that of a flying arrow, is organized. Until less than three hundred years ago there was no practical way to convert disorganized heat energy from fuel into organized energy of motion, like that of machines: 'mechanical' energy.

Accordingly, although people could use fuel-energy for many other purposes, they could not convert it into mechanical energy. If they wanted to move anything, they could use only food energy or ambient energy – the muscles of people and animals, or the structures of mills and sails. The strength of muscles is limited; water energy could be used only where the water was; wind energy was variable and unpredictable. However, by the end of the seventeenth century an idea was taking shape that eventually changed the face of the world. It was called the steam engine. In the steam engine, you use fuel energy to boil water, turning it into steam and increasing its volume a thousandfold. You then condense the steam back to water, with a corresponding decrease of volume. The changes of volume of this 'working fluid' move a piston back and forth in a cylinder, converting some of the disorganized heat energy from the fuel into organized mechanical energy. The steam engine thus overcame the crucial limitation on human use of fuel. By turning fuel energy into mechanical energy, the steam engine changed the course of human history.

By the time the steam engine was in its infancy, another major development had occurred. The use of wood for fuel had become increasingly common in Europe, especially in what is now Britain. The demand for firewood led to the gradual destruction of the great forests. By the seventeenth century, supplies of firewood were becoming scarce in Britain. People had long known that the black rocks found on North Sea beaches would burn if placed on wood fires; but even in the thirteenth century the resulting thick smoke and stench had led the governments of the day to ban the burning of such 'sea-coals' – perhaps the world's first air-pollution control measure. However, the firewood shortage in Britain triggered a move to the use of coal despite its drawbacks. Digging coal from the ground involved keeping the coal mine from being flooded by ground-water. Ironically enough the first major use of the steam engine was to drive pumps to keep the new deep coal mines from flooding: the steam engine made it easier to gather fuel to run the steam engine.

DISTURBING ENERGY

The coal-burning steam engine became a crucial factor in the 'Industrial Revolution', which dramatically transformed production methods, population patterns and social organization. Many comment-ators have offered historical evaluations of the Industrial Revolution, its social, political and ethical debits and credits. One consequence of the Industrial Revolution has, however, received comparatively little attention. Fuel-use not only increased steadily but – as we have seen – shifted from firewood, animal dung, tallow and the like, to coal, and thereafter to petroleum and natural gas. Firewood and other fuels derived directly from animals and plants contain sunlight energy that has been stored in the material for at most a century or so. When such fuels are burned they release this stored sunlight energy again, returning it to the earth's global systems from which the energy will duly be reradiated into space. The entire process of storing sunlight energy, burning the fuel to release it again, and reradiating it into space, takes place over a reasonably brief period and keeps the earth's 'energy budget' in balance. Storing sunlight energy in green plants also entails extracting carbon dioxide from the atmosphere, and incorporating the carbon into solid plant material like starches and sugars. When the plant material is burned, the stored carbon is turned again into carbon dioxide and returned to the atmosphere, thus also balancing the earth's carbon dioxide budget.

Coal, petroleum and natural gas, by contrast, are called 'fossil fuels'. Coal is the mineral or 'fossilized' remains of jungle vegetation that flourished on earth more than 150 million years ago. Petroleum is the fossilized remains of small marine creatures that thrived in the oceans at about the same time. Natural gas was formed by the decaying vegetable and animal matter as it was turning into coal and petroleum. Accordingly, the energy contained in these fossil fuels arrived on earth as sunlight not decades but hundreds of millions of years ago. Since its arrival the average temperature of the earth has undergone a number of major swings, and even created several 'ice ages'. Burning fossil fuels, and releasing this 'fossil sunlight' hundreds of millions of years after its arrival on earth, introduced a new factor into the energy balance of the planet – a time lag on a scale far longer than any associated with firewood and similar fuels.

From the mid-1940s onwards, the development of 'nuclear energy' introduced a further complication. Nuclear energy is released from the innermost parts of the atoms of certain metals, in particular uranium. Uranium is thus another type of fuel, a 'nuclear fuel'. Nuclear energy, like 'fossil sunlight' from fossil fuels, is out of balance with the earth's current 'energy budget' of sunlight arriving and 'earthlight' leaving. Indeed nuclear energy is in a category of its own: unlike 'fossil sunlight', nuclear energy has never previously passed through the earth's physical or biological systems.

Until the significant use of fossil fuels, the average temperature of the earth had been established for millenia by the balance between sunlight arriving and 'earthlight' leaving, as mentioned earlier. However, the mobilization of 'fossil sunlight' in increasing quantity has for more than a century been injecting a gradually growing amount of energy into the earth's conversion processes, in addition to the energy arriving daily from the sun. Averaged over the entire earth, to be sure, the 'fossil sunlight' contribution is still minute, and the nuclear contribution much smaller still. However, local side-effects of fuel-use are long since all too familiar: 'heat islands' over major cities regularly create anomalous weather phenomena, and aggravate air-quality problems caused by noxious products from burning fuels.

Whether fuel-use is creating global side-effects is harder to establish with certainty. One global consequence of using fossil fuels is, however, clear; and its implications are ominous. Burning any fossil fuel releases its 'fossil carbon' – carbon extracted from the atmosphere and stored in solid form many millions of years ago – back into the atmosphere, as carbon dioxide. This 'fossil' carbon dioxide is an addition to that which has in recent millenia been in equilibrium with the earth's plants, animals and natural systems. Since the 1850s the average concentration of carbon dioxide in the earth's atmosphere has increased by some 25 per cent. Carbon dioxide is not of course toxic: on the contrary, it is an essential ingredient of life. Plants inhale carbon dioxide whenever sunlight is shining on their leaves, and exhale oxygen. But carbon dioxide has another attribute – one that has suddenly made it perhaps the world's most controversial gas.

In the atmosphere, carbon dioxide molecules absorb some of the 'earthlight' heat energy leaving the earth's surface; and instead of reradiating it onward into space they radiate a fraction of it back to the earth. Atmospheric carbon dioxide therefore acts rather like the glass in a greenhouse. Greenhouse glass transmits high-quality visible light energy essentially unimpeded; you can see through the glass. Once inside the greenhouse, however, this high-quality energy is degraded to low-quality heat energy. The heat tries to emerge again from the greenhouse as invisible infrared radiation; but the glass reflects some of the heat radiation back into the greenhouse, raising the temperature inside. In the same way, atmospheric carbon dioxide reflects some of the escaping heat energy back to the earth, and thereby raises the average temperature at the earth's surface.

Other gases in the atmosphere do the same, among them water vapour, methane, nitrous oxide, and the 'chlorofluorocarbons' or CFCs best known as propellants in traditional aerosol spray cans. CFCs are the most effective 'greenhouse gases', 20,000 to 30,000 times as effective as carbon dioxide, molecule for molecule; and the other 'greenhouse gases' are also contributing to the overall change in the earth's heat-balance. Nevertheless, atmospheric scientists now believe that carbon dioxide is responsible for about half the total 'greenhouse effect'. Measuring the consequent temperature rise is difficult, not least because temperatures already vary so much from place to place and through the year. But responsible scientists in many countries now accept that 'global warming' by the greenhouse effect may already be underway. Some estimates suggest that the average temperature of the earth may rise more than 5 degrees Celsius in the next fifty years. Such a rapid temperature rise would have a profound and potentially catastrophic impact on climate and weather, the habitat of wild plants and animals, agriculture, coastal zones, major cities and indeed every aspect of life on earth.

As yet, scientists are still studying the build-up of atmospheric carbon dioxide to ascertain its causes and its effects. Burning vast tracts of tropical rainforest not only turns the tree-carbon into carbon dioxide, but removes the green leaves that would inhale the carbon dioxide once again. Marine pollution like oil slicks keeps submicroscopic plants from absorbing carbon dioxide from the atmosphere above the oceans. Nevertheless the observed increase in atmospheric carbon dioxide within the past century has undoubtedly arisen largely because humans have mobilized fossil carbon by burning fossil fuels. The more fossil carbon we inject into the earth's carbon budget, the more we shift the balance toward a higher atmospheric concentration of carbon dioxide.

The earth's climate depends on systems of bewildering complexity, sensitive to even slight disturbance. If we upset these systems by injecting extra heat-energy into them from below and reflecting heat-energy back from above, no one can say with confidence what will happen, or how fast. We are conducting a vast collective global experiment, whose result is impossible to foresee – and impossible to reverse.

Official pronouncements about the future use of fuels, both fossil and nuclear, continue to presume that such use will increase substantially in the years to come. Until recently such presumptions raised doubts only about the cost and availability of the fuels and the hardware with which to use them. Now, however, scientists, politicians and other concerned people everywhere recognize that the human use of fuels has significant global side-effects. We have begun to realize, all but too late, that the earth cannot tolerate unlimited human intervention in natural systems, including energy systems. Within the present generation we fuel-using animals may have to come to terms with limits on the conversion of fossil and nuclear fuels, if the earth is to continue to tolerate human life.

3
THE RISE OF ENERGY

Picture an average comfortable Briton, circa 1800 – let's call him John Londoner. Fuels are a part of his daily routine, so familiar to him at first hand that he gives them little thought. In London's climate, in winter especially, his house would be cold, damp and draughty, were it not that almost every room has an open fireplace, with logs of wood and lumps of coal cheerily ablaze. When darkness falls Mr Londoner lights the candles and the oil lamps, burning tallow and whale oil to give his domestic interior a smoky orange glow. He does not, of course, light the lamps with a match, since chemical matches will not be invented until 1805; he uses a tinderbox or flint-and-steel, or a hot coal from the fireplace. The fireplace wood and coal are delivered to his door by the local coalman with his horse and cart; the coalman gets the coal from the docks, where it arrives by sailing boat, perhaps from Newcastle. Mr Londoner's soup is cooked on the coal-fired stove in the kitchen, and his Sunday joint roasted in its oven. He and his family get their hot water for tea, washing and an occasional bath from a kettle on the stove or a copper over the open fire in the fireplace. He and his family also have many household appliances – brooms, dusters, mixers, grinders, bellows – all worked by hand or foot. The fuel to operate the household appliances is thus the family meal, its energy-content converted by the family members.

Of course not all the fuel of importance to Mr Londoner and his family arrives on their doorstep as fuel. Some fuel – wood and coal – is burned in kilns in the Midlands, to produce the high temperatures that fire and glaze Mr Londoner's dinnerware. Some fuel is burned in furnaces in Sheffield, to smelt and fabricate the steel for Mr Londoner's cutlery. Some fuel is burned under boilers in Lancashire, to clean and process the wool for Mr Londoner's Sunday suit; and so on. But the kilns, smelters, forges and boilers that convert the fuel-energy to manufacture Mr Londoner's possessions are technically scarcely more elaborate than his kitchen stove, or indeed his open fireplace. Neither Mr Londoner nor his family are likely to foresee the changes in store. His contemporaries in Britain, Europe, North America and elsewhere are rapidly developing new ways to use fuel, which will change Mr Londoner's society almost beyond recognition.

Mr Londoner regularly sees and handles virgin fuel materials, almost unaltered from their origins: animal or vegetable fats and oils merely rendered or moulded; wood chopped and dried but still recognizably part of a tree; coal broken into fragments but otherwise direct from the underground seam. These virgin fuel materials are burnt to release their energy in equally primitive fuel-using devices. A fireplace and chimney produce a suitable draught of air for combustion, carry away the combustion products and intercept some of the heat released, to radiate it into the room and keep the temperature more or less comfortable. But a bonfire on the dining room floor would do almost as well. In 1800, a stove is just a bonfire in an iron box, a furnace just an overgrown stove, a boiler an overgrown kettle. In each case the difference is mainly one of size; the principle of operation, converting fuel-energy to create a high temperature or to boil water, is much the same.

THE POWER OF STEAM

Even before 1800, however, inventors, engineers and entrepreneurs were devising more complex ways to convert fuel energy. The original steam engines of Thomas Savery (1698) and Thomas Newcomen (1712) had been crude, ungainly structures with a voracious appetite for coal. From 1759 onwards, James Watt introduced a succession of design innovations that made the steam engine at once more compact and more complicated. Watt's improved engines could do twice as much 'useful work' – lift twice as much water, shift twice as much rock – per tonne of coal burned. It was twice as effective or 'efficient'. As we shall see, the concept of energy 'efficiency' was to become ever more important.

By 1800 Watt and his partner Matthew Boulton had a factory in Birmingham producing steam engines of every size and variety, for mines and factories. Many types of machine, for milling, spinning, weaving and other industrial processes, were already in use, powered by wind or water. Adapting these machines to steam power at once opened new vistas for industrial development. In 1807 Boulton & Watt supplied the steam engine that powered Robert Fulton's 'Clermont', the first commercially-successful 'steamship'; ironically, of course, the engine was delivered to Fulton in the US by sailing-ship. By this time design improvements had so refined the steam engine that it could be made small enough and powerful enough not only to propel a boat through water but even to propel itself: to turn wheels underneath it, and roll itself along a pair of iron rails. Several engineers raced to devise a 'steam locomotive', while entrepreneurs set up the 'railways' on which it could run. Such railways were used initially for hauling coal; for this purpose the locomotive had to convert its fuel to supply as much mechanical power as possible. However, entrepreneurs soon thought of using a locomotive to pull a carriage-load of human passengers. For this purpose the locomotive had to convert more of its fuel-energy to kinetic energy of motion; the locomotive had to be at least as fast as a team of horses. In 1829 four groups of engineers entered their locomotives in the 'Rainhill Trials' for a prize of £500. The fastest was the 'Rocket', built by George and Robert Stephenson, which could do 30 miles per hour. The name of the winning machine was apt: the chemical rocket, invented by the Chinese many centuries earlier, was the first device that could propel itself by converting its load of fuel energy into kinetic energy.

GAS LIGHTS UP

The steam engine, stationary, marine or self-propelled, converted fuel energy into mechanical energy, delivering force and movement. But other sequences of fuel-energy conversion were also taking shape. It had long been known that you could roast coal in an oven or distil it in a retort, to make the coal give off combustible fumes. Throughout the seventeenth and eighteenth centuries scientists in many parts of Europe experimented with the gases produced by cooking coal. Daring individuals are said to have lighted rooms in Durham, in France and in Belgium, by igniting jets of gas made from coal – sometimes with explosive results. In 1787 Lord Dundonald even lighted up Culross Abbey with coal gas flames. But the first really practical application of converting coal to gas for lighting was credited to William Murdoch, a Scottish engineer who worked, as it happened, for Boulton & Watt. The company sent Murdoch to erect and supervise mine-pumping machinery at Redruth in Cornwall; and there, in 1792, Murdoch lighted his house with gas distilled from coal. At the turn of the century Murdoch, now working as an assistant to Watt, persuaded his employers to manufacture gas-making plants to provide lighting for factories. He also invented a 'gas-holder', a storage facility that allowed you to use gas at a varying rate while producing it as convenient.

A German entrepreneur, Frederick Winzler, who changed his name to Winsor, introduced a further innovation. In London in 1803 he suggested that coal be converted to gas in a central plant, and the gas delivered to customers through pipelines under the streets. Winsor stirred up public interest by giving lectures and demonstrations in the Lyceum theatre. In 1807 he lit gas-lamps along more than 200 metres of Carlton House in Pall Mall, by piping gas from his own house. Winsor's proposals were not greeted with universal enthusiasm. He met with fierce opposition from the purveyors of lamp-oil, and from people worried about the hazards of this toxic, explosive material he planned to distribute. Nevertheless, with Winsor's efforts the National Heat &

Light Company raised £1,000,000 in capital; in 1812, as the Gas Light and Coke Company, it became London's first gas company, with a charter to supply gas in the cities of London and Westminster. In December 1813 the company's chief engineer lit the new gas lamps on Westminster Bridge.

In June 1814, to mark a royal visit, a dramatic display of gas lighting was set up in St James's Park, including a wooden pagoda 25 metres high illuminated by more than 10,000 gas-lights. The display, alas, came to an unfortunate end when the pagoda caught fire and burned to the ground. But the advantages of gas-lights and well-lighted streets overcame the bad publicity from such mishaps. By 1819 several public buildings in London even had gas-lit interiors; and 460 kilometres of gas mains snaked under London streets, feeding 51,000 gas street-lights. In the US the city of Baltimore gave the lead, lighting its streets with gas from 1817 onwards. In 1819 both Paris and Brussels did likewise. By the 1850s coal gas was being produced and used in most countries in Europe, in Australia, Brazil, Canada, Mexico and India and all over the US. In Britain alone so many cities had followed London's lead that the country had over 700 gas companies.

All this industrial and commercial activity stemmed from one fact: solid coal cannot be burned conveniently as a source of light. When coal is burnt in a open fire, most of the fuel-energy released is converted into heat, and only a small proportion into visible light. Imagine the difficulty of installing a bracket to burn coal high enough up the wall of a room to provide useful illumination, and of keeping the coal-fire alight without suffocating everyone in the room with the accompanying deluge of heat. It therefore made sense to Murdoch, Winsor and many others to process the coal, to transform it from a 'primary' virgin fuel with undesirable characteristics into a 'secondary' fuel whose characteristics were more desirable.

Murdoch's coal-distilling retort was thus an intermediate stage, converting one fuel into another. The 'end-use' conversion took place in the flame of the gas light. As it happens, coal gas burns with a clean flame that contains scarcely any solid particles to provide 'incandescence', the effect of the glowing soot that makes a candle flame shine. Accordingly, gas lights were comparatively convenient; but they were not in fact very bright. Not until 1885 did Carl Auer von Welsbach invent the 'gas mantle' that bears his name. The Welsbach mantle is a delicate filigree of thorium and cerium oxides, which absorb the heat-energy of the gas flame and reradiate it as visible light energy, blue-white and brilliant, familiar today in campers' lamps. But until the invention of the Welsbach mantle the end-use conversion of coal gas was directly analogous to Mr Londoner's open coal fire, or indeed a wood-burning bonfire. When the fuel was simply burned in the open, the user had to be satisfied with whatever combination of energy-conversion resulted, an unpredictable mixture of heat, light, mechanical motion of the air, possibly even a low whistle. To be sure, engineers worked toward better control of combustion, especially to reduce the loss of potentially useful heat; but manipulation of energy-conversion at the point of use was in its infancy. Among domestic uses of fuel-energy, for instance, only lighting received such attention. Cooking and heating still involved burning coal directly, in the immediate vicinity of the food or the furniture. However, more subtle energy conversion processes, especially at the point of use, were on their way. They received a powerful stimulus from an unexpected direction.

ELECTRICITY IN ACTION

The phenomenon of 'electricity' had been known since Biblical times. The word itself comes, improbably, from the Greek word 'elektron', meaning amber – because amber, if rubbed, would attract bits of cork or other light material. Such 'static electricity' remained for millenia a mere curiosity. However, in the eighteenth century, inventors devised 'friction machines' that could produce enough static electricity to make sparks jump. Luigi Galvani, an anatomist in Bologna, thought that such a friction machine also made the leg of a dead frog on his dissecting table jump. But his countryman Alessandro Volta proved in 1800 that the electrical impulse came from placing two different metals in acid; and Volta built a 'Voltaic pile' that would deliver a steady flow of electricity. Electricity became one of the many scientific parlour tricks popular at the time. Guests would stand in a ring, holding hands. At a signal the two guests at either end of the ring would grasp with their free hands the two terminals of a Voltaic pile, and everyone in the ring would get a sharp shock as an 'electric current' flowed through their linked hands. But the phenomenon seemed to have limited practical promise. It was not yet recognized as a manifestation of another kind of energy – electrical energy – produced by converting the chemical energy stored in the 'fuel' – that is, the metal plates of the Voltaic pile.

Descendents of the Voltaic pile, called 'electric batteries', in due course proved to have important uses; but the basic concept has shortcomings still troublesome two centuries after Galvani and Volta. Constructing a compact system able to store a substantial quantity of chemical energy and release it as electrical energy was and still is difficult. Producing a compact store of chemical energy is easy enough; but the result is likely to be a package that will release its energy almost instantaneously, like a stick of dynamite. A package that will release its chemical energy slowly, and as electrical energy, is likely to be very bulky and expensive.

Nevertheless, despite the shortcomings of the first electric batteries, they opened a whole new avenue of scientific research. In 1820 a Danish scientist named Hans Christian Oersted made an extraordinary discovery: when an electric current flowed through a wire, the needle of a nearby magnetic compass would move. Conversely, if the magnet was fixed in place, the wire itself would move. In other words, electrical energy could be converted directly into mechanical energy of motion. By 1831 the British scientist Michael Faraday had demonstrated a yet more extraordinary corollary: if you moved a magnet near a wire, or a wire near a magnet, an electric current flowed in the wire. In other words, mechanical energy of motion could be converted directly into electrical energy.

Widespread practical applications of these intriguing possibilities, however, occurred only gradually. Within the 1830s small 'electric generators' based on Faraday's discoveries were on sale. Such a generator consisted essentially of a coil of wire that could be rotated, by hand or machine, between the poles of a magnet; the coil wires led to terminals from which you could draw electric current. Larger generators, whose coils were turned by waterwheels or steam engines, delivered a more substantial flow of current, and led to the first really practical use of electricity.

Faraday's scientific colleague and mentor, Humphrey Davy, had discovered in 1810 that when the terminals of a large battery were connected to two carbon rods placed close together, an electric current would flow across the gap between the rods, like a continuous spark. The result was a brilliant incandescence from the minute carbon particles in the hot gap: an 'electric arc'. There was no such thing as a soft arc light. If the arc was formed at all its garish glare was utterly unsuitable for interior lighting; and battery power was unsuitable for outdoor lighting, because of the short service life and cost of the battery. However, powering an arc light with an electric generator was much more feasible. By 1857 you could visit a British lighthouse in which coal was burnt to raise steam to run a steam engine to turn a generator to power an arc light. The lighthouse keeper, stoking the boiler, was converting chemical fuel energy to heat energy to mechanical energy to electrical energy to light energy, all for a special purpose: to keep mariners off the rocks.

Improved designs of generator, using 'electromagnets', became available in the 1860s; and major manufacturers – Siemens from Germany, Brush from the US and Crompton from Britain – were soon offering complete arc-lighting sets on a growing international market. By 1878 arc lights were being used in theatres and factories; at the Bramall Lane football ground in Sheffield 30,000 spectators watched an evening match under arc lights, whose strange brightness reportedly provoked odd mistakes on the playing-field. In 1879 Blackpool became the first seaside town to boast 'illuminations'. The growing popularity of electric arc lighting sowed doubts in the minds of the gas companies and their shareholders, who feared an upsurge of competition. But a committee appointed by the main London gas company reported reassuringly in 1878 that electric light – that is, arc light – 'can never be applied indoors without the production of an offensive smell which undoubtedly causes headaches, and in its naked state it can never be used in rooms of even large size without damage to sight . . .'

On 18 December 1878 a Newcastle chemist named Joseph Swan pricked the gas companies' smug bubble. At a meeting of the local chemical society Swan showed that a carbon fibre inside an evacuated glass globe would glow when carrying an electric current. Swan had invented the incandescent 'light bulb'. The light it produced was gentle and pleasing, and it worked without noise or odour. In 1879 an American inventor named Thomas Edison patented a remarkably similar invention, and acrimony ensued; but in 1882 the two inventors joined forces to form the Edison and Swan United Electric Light Company. By this time indoor electric light had become a luxury status symbol. By November 1880 Lord Salisbury's stately home was lighted with Swan lamps powered by an electric generator driven by an adapted sawmill; a month later Sir William Armstrong's home was likewise illuminated, with electricity from a water-driven generator. Within the next five years many major buildings – the House of Commons, the Savoy Theatre, the British Museum, the Royal Academy and others – were electrically lighted. Each building, of course, had to have its own expensive coal-fired steam-driven generating plant; but the extra maintenance this entailed was offset by the ease of maintaining and controlling the lamps themselves, and by the quality of the light, compared with paraffin lamps, gas lamps or arc lights. The new form of useful energy, electricity, had arrived to stay.

In the early 1880s, as electricity gained rapidly in public appeal, the entrepreneurs moved in. Soon they were setting up companies not only to supply arc lights and incandescent lights but also to offer a supply of electricity itself, to save customers the cost of installing their own generating equipment. Public electricity supply, like public water supply or main drainage, is a natural monopoly; to have two or more competing electricity suppliers laying cables and fighting for customers in the same street makes no technical nor economic sense. Accordingly, the Electric Lighting Act of 1882 required that private electricity suppliers who wished to lay underground cables had to apply for an exclusive franchise under the Act, and established legal control over electricity supply.

Despite the commercial optimism of some electricity suppliers, in Britain electric lighting was still trailing far behind gas lighting in popularity. Electric lights were still significantly more expensive to operate than gas lights, and the Welsbach mantle had improved dramatically the quality of light from gas. One further and more fundamental problem also troubled electricity suppliers, and made electricity more expensive. Understandably enough, customers mostly wanted artificial light when it would otherwise be dark – in practice the hours of evening. A gas supplier could cope by running the gas-making retorts more or less continuously, and feeding the gas into huge expanding storage tanks, commonly albeit inaccurately called 'gasometers'. When twilight fell and customers began to open their gas jets and light up, the gas stored in the gasometers could be fed into the gas mains as required, until bedtime and lights out. The gasometers could be refilled gradually through the small hours and the following day.

Unfortunately for electricity suppliers, however, there was – and is – no way to store electricity in 'bulk' until it is required. Electricity is not a fuel; it is an 'energy carrier', but only in the sense that it is a manifestation of an energy conversion process actually occurring, moment by moment. Electricity has to be generated in precisely the quantity being used, as it is being used. This entails a high degree of operating finesse. It also means that an electricity supplier who wants to keep customers satisfied must be able to call on enough generating capacity to supply the largest quantity of electricity the customers may collectively desire to use at any given instant: the 'peak load' on the system. In 1895 an electricity supplier had to have enough generating plant to meet the 'peak load' at evening mealtime or thereabouts, when customers would have most of their lights on. Much of this generating plant was however required for only two or three hours each evening, and lay idle throughout the rest of the day and night. So long as an electricity supplier could produce saleable electricity from his costly plant for only a few hours a day, he would be at a serious competitive disadvantage. The electricity supplier could not reschedule the onset of nightfall; his only recourse was to find other possible uses for electricity at other times of the day.

The search did not take long. An electric generator, converting mechanical energy into electrical energy, can readily be turned back to front, converting electrical energy into mechanical energy and becoming an 'electric motor'. Engineers had recognized the possibility from the time of Faraday. For several decades, however, they had seen little justification for starting with fuel, converting it to mechanical energy in a steam engine, then to electrical energy to send through a cable, then back to mechanical energy by means of a motor at the other end of the cable – losing significant amounts of energy as heat at every stage. Why not merely take the fuel itself to the point of use and convert its chemical energy to the required mechanical energy on the spot, using a steam engine? The energy required to carry the fuel by horse and cart to the point of use instead of sending its energy electrically, a few miles at most, would be far less than the energy lost in all the conversions to electricity and back. Gradually, however, one particular application of the electric motor began to look promising.

By the 1880s, steam-powered railways were criss-crossing Britain and linking many towns and cities, carrying both freight and passengers. Within the towns and cities, however, almost the only transport available was still horse-drawn: private horse-drawn carriages and carts running on the streets, and horse-drawn 'trams' running on rails sunk into the streets, providing public transport for fare-paying passengers. Some attempts were being made to build actual railways within cities, notably the Metropolitan Line in London. But the steam locomotives, belching smoke and cinders, were far from welcome at street level, and made travel through underground tunnels a penance for passengers and crew alike. An electric motor, by contrast, ran without any noxious emissions whatever; the noxious emissions could be released somewhere else, at a stationary coal-fired steam-power electricity-generating plant, sending electricity along overhead cables for a mile or more to run the motor. In 1892 the Bradford town tramway was converted from horsepower to electricity, increasing the speed of the trams and more than doubling the number of passengers each could carry. Thus, although the electric light was still outshone by gas light, the electric motor had demonstrable advantages over its main competitor, the horse. Electrification of urban and suburban tramways proceeded apace. Using the generating plant both for trams in the daytime and for lights in the evening would be much more economic.

Using electric motors in factories would improve the economics of the system yet further. Electric motors, comparatively clean and flexible, were beginning to make inroads on the conventional use of on-site steam engines. A single electricity generating plant could power a whole array of electric motor-driven machines in different parts of a factory, and allow them to be independently controlled, started and stopped. Powering the machines directly by steam would require either a large number of separate steam engines – inefficient, noisy and noxious – or a complicated linkage of shafts and belts from a single steam engine. By the turn of the century factory inspectors were remarking on the growing popularity of electricity in factories, for power as well as for lighting. Much of this electricity was generated on factory sites, by privately owned plant; but an increasing proportion was purchased from public electricity suppliers. The public suppliers thereby at last achieved the improved utilization of their plant that had eluded them as long as they supplied only lighting. Factories operated not only during the day but often around the clock; the sale of electricity at all hours enabled the public suppliers to run their plants more economically, to recover their costs more effectively, and to reduce their charges per unit of electricity. From 1900 onward the role of electricity as an energy carrier expanded steadily, throughout the whole of industrial society.

By this time it was evident that a wide range of human activities could benefit from the imaginative use of fuel energy, by employing appropriate hardware. In each case the fuel was the same: what differed was the hardware to convert it for use. In 1900 coal was much the most important fuel in Britain and many other industrial countries. Starting with a given wagonload of coal you could accomplish many different objectives, depending on the hardware you used to convert the coal's energy: fireplaces, furnaces, boilers, steam engines, gas retorts, gas jets, electricity generators, light bulbs, motors and so on. The historical sequence is significant. Coal itself came first. Over decades and centuries it was subjected to progressively more extended conversion processes, to accomplish some specific human objective: warmth, high temperature, hot water, steam, mechanical motion and power, light, even sound. At every stage human ingenuity focused on the hardware used to convert the energy. Engineers sought to make it more effective, to produce suitably bright, pleasing light, or reliable controllable temperature, or smooth mechanical movement – whatever the user desired. Engineers also sought to make the hardware both less expensive and more efficient.

Throughout the nineteenth century, in pursuit of such specialization, engineering design became more and more specific. For instance, electricity generators operated best when the coil of wire between the magnets moved in a circle. Conventional steam engines, however, moved pistons back and forth. To be sure, you could convert this motion into rotation by using cranks and linkages; but such arrangements lost a lot of energy as frictional heat and were in any case inelegant. Could you not convert the mechanical energy of expanding steam directly into rotating motion? The analogy of the windmill yielded the answer. In 1884 a young British engineer named Charles Parsons patented a sort of 'steam windmill', which he called a 'turbine', from the Latin word for 'spinning–top'. Parsons' steam turbine was basically a single shaft along which were mounted a series of multibladed 'windmills', each slightly larger than its precursor, with the whole arrangement inside a contoured pressure-chamber. Hot high-pressure steam from a boiler was directed onto the smallest windmill; as the steam expanded along the chamber it passed through each windmill in succession. The increasing size of the windmills exactly compensated for the expansion and the decreasing pressure of the steam. All the windmills acting simultaneously spun the shaft on which they were mounted, and the shaft spun the coil between the magnets. The steam turbine 'revolutionized', in both senses, the generation of electricity from fuel, greatly increasing the amount of electricity you could produce from a given amount of fuel. By the early 1900s most of the new and larger fuel-burning electricity-generating plants were using steam turbines.

The growing popularity of electricity also gave a fresh impetus to the old watermill. A traditional watermill, converting the energy of falling water directly into mechanical energy to turn a millstone or other machinery, could be used only where you had a supply of falling water. However, if a water wheel turned not machinery but an electrical generator, the electricity could be delivered some distance away, and converted back to mechanical energy in an electric motor. Indeed you could even tap major waterfalls, which would spin a millwheel so fast it would disintegrate. You could also create an artificial waterfall by damming a river to produce at one location a sudden precipitous drop. Water falling over this drop could be fed through a waterwheel of more sophisticated design – possibly a so-called 'water turbine' akin to the steam turbine. The water turbine could turn a generator to feed electricity into cables to be carried to users far from the river. Countries with usable falling water, notably Canada, Norway and Sweden, soon began to establish a whole new type of electricity supply called 'hydroelectricity'.

OIL, PETROL AND INTERNAL COMBUSTION

In Britain, hydroelectric development was confined to small installations in remote locations. Throughout the nineteenth century and into the twentieth, Britain's expanding industrial economy was based on fuel energy; and much the most important fuel was coal. The same was broadly true of most other countries with expanding industries. In the US, however, another major development was underway, that would have a world-shaking impact. To be sure, its beginning was hardly impressive. In 1859, near Titusville in Pennsylvania, a small-time entrepreneur with a spurious military title, 'Colonel' Edwin Drake, dug a well. He was not, however, looking for water. He was looking for oil; and he found it. The timing of his discovery was propitious. Although US towns and cities had long since brightened their streets with gas light, the great majority of houses and other buildings still relied on candles and whale oil for lighting; and whale oil was becoming both scarce and expensive. The market for the new 'rock oil', or 'petroleum', was primed. Within five years the oil drillers had thoroughly perforated western Pennsylvania, and were spreading across the country.

Petroleum itself proved to be a cornucopia. Because it was liquid, it could be heated easily to boiling-point – or rather to an entire succession of boiling-points: petroleum or 'crude oil' was a complex mixture of many different substances that could be boiled off selectively and recovered separately. At first this 'refining' only divided the petroleum into 'kerosene' for lamps and stoves, and lubricating oil for machines – the only major uses for which customers existed. The rest was poured away. Gradually, however, oilmen found it worthwhile to subdivide a barrel of petroleum into a dozen or more 'fractions', each with specialized uses.

The oil had to be transported from wells to refineries and thence to users; and the railroads had recently laid track from Cleveland through western Pennsylvania. One man in particular divined the commercial potential of a disciplined, if not indeed rigged, organization of oil production, transportation, refining and marketing. His name was John D. Rockefeller. Rockefeller left the drilling to others. His fastidious and distant nature recoiled from the chaotic melee of the drilling scene. Drillers were so recklessly enthusiastic that they regularly put themselves out of business, pouring so much oil onto the market that its price fell from US$20 a barrel in 1860 to 10 cents a barrel in 1861. As gluts kept recurring, a barrel of oil was sometimes literally cheaper than a barrel of water; and many a driller was wiped out by debt and despair. Rockefeller wanted no part of such foolishness. Quietly, deviously, he schemed and bought his way into commanding control of the oil business, not by producing oil but by taking over refineries, fixing prices and arranging secret cut-price contracts with railroads. By 1870, after only seven years in the business, Rockefeller could set up a company called Standard Oil, which already held one-tenth of the entire US oil industry.

By 1875 Rockefeller was so firmly entrenched at the top of the oil industry, with his railroad agreements and his refining cartel, that he could make other refiners offers they could not refuse. By 1883 his Standard Oil Trust extended right across the US. Already the railroads were seeing the ironic aftermath of their early involvement with Rockefeller. Cut-rate rail transport had given Rockefeller his first leverage against his competitors. By 1883, however, the railroads had lost much of their oil traffic to Rockefeller's new 'pipelines', which by that time criss-crossed much of the eastern US. For the railroads, worse was to come. By the 1880s, oil was on the threshold of symbiosis with a new way to convert fuel energy: the 'internal combustion' engine.

In the 1880s, in the US and Europe, the horse still ruled the roads. In the early 1800s, to be sure, entrepreneurs had attempted to introduce coal-fired steam-powered road vehicles. But the coach proprietors organized public opposition, and railway interests with their influence in Parliament did their best to nobble this potential competition before it got under way. Their most notorious move was to persuade Parliament to pass, in 1836, a law stipulating that any steam-powered road vehicle be preceded by a man carrying a red flag to warn the populace and the horse-drawn traffic. This measure effectively thwarted the fuel-powered car for decades; not until 1896 was it repealed.

In the twentieth century, nevertheless, the car has revenged itself comprehensively on the railway – albeit not by way of the steam engine. By the mid-1800s engineers were seeking to devise an engine that could convert the fuel energy of coal gas directly into mechanical energy. The fuel to run such an engine could be delivered to the site through gas mains, much more easily than cartloads of coal. However, converting coal to gas, piping the gas to the site and then burning it in a steam engine would waste too much of the original fuel energy. The key to the problem was that, unlike coal, gas would not only burn steadily but would also, under suitable conditions, explode – releasing all its fuel energy abruptly and violently.

For lighting, of course, the explosive potential of coal gas was an unwelcome embarrassment with occasional unfortunate consequences. Nevertheless, if a small amount of gas could be blended with air into an explosive mixture, and ignited in a confined space like the cylinder of a steam engine, the explosive release of fuel energy would push the piston much like the expanding steam in a steam engine. In such a 'gas engine' the fuel would burn in the cylinder itself, eliminating the steam-raising stage: an 'internal combustion' engine.

From 1860 onwards, various designs of internal combustion engine were built, albeit with limited success. The breakthrough came in Germany, in 1876, when August Otto patented a novel and inspired principle. Otto injected the gas-air mixture into the cylinder while the piston was withdrawn, so that the piston, returning into the cylinder, compressed the explosive mixture before it was ignited. The fuel energy was therefore converted to mechanical energy at a substantially higher temperature and pressure, making Otto's engine much more compact and efficient than its precursors. The internal combustion engine was off and running.

Otto's engine and its immediate successors were of course intended for stationary operation; you could not deliver coal gas to a moving target. In 1885, however, Otto's countryman Gottlieb Daimler incorporated Otto's design principles into a compact single-cylinder engine fuelled not by coal gas but by a volatile fraction of petroleum called 'petrol', which evaporated readily to form an explosive mixture to inject into the cylinder. In 1886 Daimler attached his petrol engine to a bicycle, to produce the first 'motorbike'. Petrol, as a liquid, could be carried in a tank on the bicycle and pumped through a pipe into the engine. Furthermore, petrol packed a substantial amount of recoverable chemical energy into a small tankful. You could carry enough stored energy on the bike to propel it a considerable distance; at the time, of course, you could not purchase more petrol at a nearby streetcorner. By powering a bicycle with a petrol engine, Daimler underlined the advantage of using an internal combustion engine burning liquid fuel. No one had ever proposed a coal-fired steam-powered bicycle; the fuel-handling equipment alone would have been much too cumbersome.

The ensuing decade saw the emergence of the 'motor-car' in Germany, France, Belgium, Britain, the US and elsewhere. At first it was a novelty for aficionados. But its requirement for liquid fuel from a different fraction of the refined barrel of oil presented an alluring and congenial consort for the oil industry. Once the two found each other, there was no stopping them. Throughout the twentieth century no technological marriage has been more potent that that of oil and the car.

The petrol engine, able to deliver, weight for weight, far more power than the strongest horse, also helped human beings to realize an age-old dream. Two hard-headed and dedicated brothers who owned a bicycle shop in Dayton, Ohio, finally succeeded where so many precursors had failed. In 1903, amid the dunes of Kitty Hawk, North Carolina, Orville and Wilbur Wright proved that by using the stored energy of petrol, converted by an internal combustion engine carried on artificial wings, a human being could fly, in a machine 'heavier than air' – an 'aeroplane'.

ENERGY AND INFORMATION

By converting fuel energy in locomotives, trams, cars, steamships and aeroplanes, you could move people and goods ever farther and faster. But a parallel development was also taking shape, that would in due course present a major alternative to transport and travel. Signalling from one place to another had of course been common since prehistoric times – think of beacon-fires on hilltops, flags on sailing-ships, hand-written letters carried from writer to recipient. However, long-distance transmission of information was limited in range, speed or vocabulary – usually all three. The discoveries of Volta, Oersted and Faraday – how to produce electric current and use it to deflect a magnet – did not, as already mentioned, lead at once to large-scale applications. But they led with remarkable speed to another type of application – one that converted comparatively little energy, but with increasing subtlety.

In 1836, within five years of Faraday's ground-breaking work with electricity and magnets, an American named Samuel Morse had invented a system of electromagnetic hardware called 'telegraphy' – 'writing at a distance'. The essence of the 'telegraph' was simplicity itself. The sender of a message had a switch that could close or open – 'make or break' – an electrical connection, starting or stopping the flow of current from a battery. At some other point on the wires, which could be many miles away if the battery was powerful enough, starting or stopping the electric current would activate or shut off an electromagnet, working a buzzer or inking a mark on paper. Morse devised a concise 'code' of short and long buzzes or marks, called 'dots' and 'dashes', to represent letters of the alphabet and other symbols. The most famous message in Morse code is 'dot dot dot pause dash dash dash pause dot dot dot': S O S.

Morse's system received a substantial boost in 1842, when the US Congress granted $30,000 for an experimental telegraph line linking Washington DC and Baltimore. Its success triggered the rapid spread of Morse telegraphy, not only in the US but internationally. Within twenty years the combined potential of the steamship and the telegraph were being applied to a project that even a century later excites awe: laying a telegraph cable from Europe to America, on the floor of the Atlantic Ocean. In 1860 the world's largest steamship, the Great Eastern, designed by the legendary British engineer Isambard Kingdom Brunel, made the fastest crossing to date from Britain to New York: eleven days – the most rapid communication then possible between the two continents. In 1869 the Great Eastern joined other ships in laying the transatlantic telegraph cables, and thereby helped to make the inter-continental communication of information almost instantaneous – a few minutes for a long message. Within the space of a decade the electromagnetic telegraph had made the world much smaller.

The subtle conversion of energy to transmit information was nevertheless still in its infancy. Telegraphy, as its name indicates, could transmit information only as writing or spelling. However, in the early 1870s Alexander Graham Bell, a Scottish professor of vocal physiology at Boston University in Massachusetts, combined his knowledge of voice-production with ingenious hardware to convert the energy of sound in air into electrical energy, and then back into sound energy. One afternoon Bell, working in one room of his laboratory, spoke into a curious instrument the less-than-immortal words 'Mr Watson, come here; I want you'. In another room his assistant heard a metallic reproduction of his boss's command, and complied. The 'telephone' – 'sound at a distance' – was born. In 1876 Bell patented his invention; by 1900 the 'phone' was ringing in homes and offices throughout Europe and North America.

However, the potential for subtlety in converting energy to convey information was far from exhausted. In 1864 a British scientist named James Clerk Maxwell published a landmark treatise in which he predicted that under suitable conditions an electric current would send out waves of energy, exactly analogous to light waves but invisible. A German scientist named Heinrich Hertz began experiments seeking these 'electromagnetic waves'. In 1886 Hertz set up in his laboratory a battery-powered electric circuit with a gap in it, across which he could make a spark jump. On the other side of the lab was a similar electric circuit incorporating a sort of electromagnet, but no battery or other energy source. Hertz found that when he made a spark jump in the battery-powered circuit, a similar spark jumped in the other circuit. Energy was being transmitted from one circuit to the other, although they appeared unconnected. Hertz had found Maxwell's electromagnetic waves. For some years they were called Hertzian waves; but gradually they become known as 'radio waves'. Hertz was not, however, denied his place in history; the unit of 'frequency', the number of cycles of a wave per second, is now officially named the 'Hertz', and we remember him in 'kiloHertz' and 'megaHertz' when tuning the radio.

Hertz's waves soon attracted the engineers, led by an astonishingly precocious young Italian named Guglielmo Marconi. While still in his early twenties Marconi became convinced that radio waves could be used to carry information without the medium of a cable: 'wireless telegraphy'. Shortly thereafter he succeeded in sending telegraph signals between Penarth in Wales and Weston in England without wires; and in 1899, the year of his twenty-fifth birthday, he established an operating 'wireless' link across the Channel between England and France. Not content with this, Marconi went on to show, in 1901, that he could send a wireless signal 2100 miles, from Cornwall to St John's, Newfoundland.

Travel and transport were thus joined by 'telecommunications' – and human control of energy conversion grew ever more elaborate and elegant. You could convey energy in fuel to wherever you wanted to use it, store it there and convert it as desired. You could convey energy by electricity. Or you could convey information: energy so carefully organized that converting it could deliver an intelligible meaning, even on the other side of the world.

A CHANGED WORLD

By 1914, John Londoner's grandson Edward lived in a world his grandfather would scarcely have recognized. Edward Londoner could turn night into day with the flick of a switch. He could make his surroundings warmer, or indeed cooler, at will. He could exert more force than had ever been available from slaves and draught animals. He could move farther, faster and higher than any animal could carry him. He could send and receive messages all over the world, almost literally as fast as light. Using fuel energy and the hardware to convert it, Edward and his contemporaries could create and control ever more extreme physical conditions, of temperature, pressure, speed, height, depth, size and strength. Using fuel had in turn wrought startling changes in society. Edward lived neither in the country nor, strictly speaking, in the town. He lived in a 'suburb', some distance from where he worked. He travelled to and from work by tram. He worked at night, in a factory, operating machinery run by an electric motor, making car parts. Grandfather would have found the whole way of life incomprehensible; but grandson took it for granted.

Changes in individual lives were accompanied by those in industry and commerce. In 1800 the 'energy business' consisted essentially of wood-cutting and coal-mining, wholesale and retail coal and wood merchants, suppliers of whale oil and candles, and manufacturers of basic conversion hardware like grates, stoves and of course the new steam engines. By 1914 the original fuel suppliers had been joined not only by the petroleum suppliers but also by major new industries converting raw virgin fuel into specialized forms for specialized purposes: coal into gas and electricity, and petroleum into petrol, fuel oil and other derivatives. The manufacture of energy hardware was a large-scale industrial sector in its own right, turning out everything from turbines, generators and motors to cars and buses to household appliances – fans, vacuum cleaners, even 'electric fires'.

The fuel suppliers themselves, led by the petroleum industry, had embarked on expansion, mergers and high-level integration, gradually transforming themselves into corporate enterprises of unheard-of size and influence – and transforming the ground-rules of commerce, from a simple marketplace into an intricate and near-impenetrable labyrinth.

By 1914, using fuel energy had given humans the opportunity to confront and overwhelm their 'natural' surroundings, at least temporarily. Unfortunately, using fuel energy had also given them the opportunity to confront and overwhelm each other, with unprecedented variety and violence. Despite their accelerating technical imagination and skill, people themselves appeared to have changed little; and age-old human proclivities combined to plunge most industrial areas of the world into a brutal and ugly war. Fuel energy and fuel-using hardware let people kill each other faster, farther away, higher, deeper, more indirectly and more indiscriminately than ever before. Motor vehicles, motor vessels, aircraft, radiotelegraphy and other marvels of modern energy hardware were pressed into military service, with dramatic effect not only on the course of hostilities but also on the development of the hardware.

The first 'World War', while killing millions of combatants and civilians, also gave great impetus to technological advance. The paradox is grim, and more intractable some eight decades later. It also illustrates vividly the dichotomy underlying the use of fuel from prehistoric times onwards. Using fuel has never been an unqualified blessing to humanity. Every new application has both advantages and disadvantages, to individuals and to society as a whole. In some ways, for some people, the Industrial Revolution was a spectacular manifestation of human 'progress'. In other ways, and for other people – perhaps the majority – its immediate benefits were debatable, if not indeed non-existent. The same troubling questions about the role of technology in human society have persisted ever since. In the final decade of the twentieth century they have acquired a fresh urgency.

From 1918 the use of fuels expanded rapidly throughout industrial society. The use of electricity not only expanded but diversified, from lights and motors into activities that only electricity could accomplish. 'Electrochemistry' found major industrial applications, notably in smelting aluminium; once an exotic rarity, aluminium became not only commonplace but eventually trivial, in beer cans, and TV dinners cooked by 'microwaves' – like radio waves but shorter. Manipulating electricity with ever more subtlety gave rise to a new engineering discipline called 'electronics'. Wireless telegraphy and telephony combined to produce 'wireless telephony' or, as it has long been known, 'radio', followed in 1936 by 'television', and in the 1980s by 'telefacsimile' – 'fax'. Together they gradually enveloped the planet in a sea of artificial electromagnetic energy carrying human information – everything from distress calls to singing commercials. Electronics created a whole catalogue of characteristic hardware, designed to convert energy with extraordinary precision and delicacy. 'Vacuum valves' gave way to 'transistors', and thence to 'microchips'. Calculating machines became computers; computers shrank, sometimes down to pocket-size; and computers learned to use the telephone, immersing the earth in a global flux of electronic information.

Nor was innovation limited to electricity. In the 1930s the petroleum industry, especially in the US, at last devised a way to use a product that had hitherto been not only a nuisance but a real danger. Underground rock formations bearing coal or oil sometimes also contained a fossil carbon-compound so light it was not liquid but gaseous. The gas had been known for centuries in coal mines, as the dreaded 'fire damp'. Many miners had died in fierce explosions triggered when a miner's lamp or spark ignited fire damp from the coal seam. Improved mine ventilation and Humphrey Davy's shielded safety lamp for miners gradually reduced the hazard of fire damp. But when the oil-drillers set to work, the same gas frequently emerged with the oil. If a spark from the drill rig ignited the gas, the column of roaring flame billowing skyward could be extinguished only by means equally violent, like dynamite. The drillers had to devise a way to separate the precious liquid oil from its troublesome gaseous accompaniment, feeding the separated gas into a tall tower topped by a continuous flame called a 'flare'. You could always spot an oilfield from the air, day or night, by its flares.

Not until the 1930s, by developing thin-walled high-pressure pipelines, could the oil companies at last turn this irksome gas into a profitable fuel. Once they had done so, however, 'natural gas' was to become increasingly desirable in its own right. Emerging under pressure from deep in the earth, it required virtually no processing; it had only to be channelled into the pipes and carried right to final users, where it could be burned in simple jets in cookers and gas-fires and even under boilers. Natural gas burns with a hot clean flame, turns completely into carbon dioxide and water vapour and leaves no residue whatever. Soon natural gas pipelines were snaking all over the US. Drillers still preferred oil; but they were no longer too disappointed when a 'wildcat' exploratory well yielded no oil but rather a rushing torrent of natural gas, invisible fuel ready for use. As an energy carrier, natural gas had many of the virtues of electricity, was much simpler to produce, and – unlike electricity – could even be stored.

In Britain and most other industrial countries, the gas industry still produced its gas from coal or sometimes oil, and was increasingly hard-pressed by electricity. Electricity could not only deliver light and heat and drive electric motors, but could also power the novel electronic devices rapidly becoming fashionable, like radio and television receivers. In desperation the gas suppliers even for a time tried to market monstrosities like a gas-operated radio; but gas from coal and oil continued to lose ground to electricity. The advent of natural gas, however, set the stage for a protracted and stubborn confrontation between the two competing energy carriers.

At the end of the 1930s war erupted once again. The first world war had introduced new military hardware and strategies based on using fuel-energy; but it could nevertheless have been fought with only traditional uses of fuel. The second world war, however, depended from its onset on ever more subtle uses of energy – and was ended by the catastrophic broaching of a store of fuel-energy hitherto inaccessible to people: the nucleus of the uranium atom. The war not only reinforced existing applications of energy-use, but swept scientists and engineers into intensive innovation: the gas turbine or 'jet engine', radar, rocketry, and – looming over all the others – the release of 'nuclear energy'. The international collaboration that created the weapons that obliterated Hiroshima and Nagasaki was the largest scientific and technical enterprise ever undertaken. It was also the most secret.

Purely as an energy device, the 'atomic bomb', as it was at first called, was nothing of the kind. A conventional bomb like the 'blockbuster', made of chemical high explosive, was truly 'atomic', since the energy released when it exploded came from the atoms of its charge. However, the energy that devastated the Japanese cities – heat, light, acoustical shock waves and radiation – came not from atoms as a whole but from the much tinier inner cores or 'nuclei' of atoms of the metals uranium and plutonium – not 'atomic' energy but 'nuclear' energy. Unlike all the 'renewable' fuels – wood, dung and such – and fossil fuels – coal, petroleum and natural gas – uranium and plutonium stored energy that had never at any time been carried to the earth by sunlight. They were in an entirely new category: 'nuclear fuels'. Indeed in some respects they were not so much conventional 'fuels' as energy hardware. Every other virgin fuel can if desired be burned simply and directly, on a bonfire or at a wellhead. Energy from nuclear fuel, however, can be used only through a long chain of complex industrial processes. Once fabricated, a piece or 'element' of nuclear fuel remains in a 'nuclear reactor' for several years pouring out energy – more like a component of the plant than like, say, a tonne of coal.

In any case, in the decade following 1945, using uranium as a fuel took second place to using it as an explosive, in nuclear weapons. Plans to use uranium for 'peaceful' purposes like generating electricity began to take shape at the very time when a competing fuel, petroleum, was about to become cheaper and more abundant than ever before. By 1950 the worst immediate effects of the second world war were fading. In

Europe and Japan industries were being rebuilt; the US had emerged as the world's most powerful nation; and the worldwide role of fuel energy was on the threshold of an extraordinary upsurge. In the decade before the second world war preliminary exploration had identified enormous reserves of petroleum around the eastern end of the Mediterranean – what came to be known as the Middle East. Amid a welter of international intrigue seven huge oil companies, abetted by Britain, France, the US and other governments, got access to this oil on very advantageous terms. The oilfields in Iran, Saudi Arabia and other Middle Eastern countries yielded their treasures readily, and exploration soon located a series of 'giant' oilfields. Cheap oil flooded the world markets – Europe, Japan and even North America, the birthplace of the oil business.

DEVELOPING DEPENDENCE

In 1959, US domestic oil-drillers persuaded the US government to impose oil-import quotas, to protect the indigenous oil industry from the cut-rate competition. In the longer term, this US policy meant using up its own oil first, and paying over the odds for it, instead of saving its own oil and buying cheaper overseas crude while it was available – illustrating vividly the influence of US indigenous oil-producers. Nevertheless the policy ran its course in only a decade. By 1970 US annual use of crude oil had outstripped annual domestic production; the US became, for the first time in history, a net importer of oil.

It had plenty of company. Almost every other western industrial country had already succumbed to the seductive temptation of petroleum at three US dollars a barrel. Rivers of oil from the Middle East, augmented by tributaries from North Africa, Nigeria, Indonesia and elsewhere, poured into western Europe, Japan and other customer countries. The cheap and copious crude brought about a dramatic shift in patterns of energy-use throughout the industrial world. The expanding refineries decanted petrol for cars, heavier 'diesel' oil for heavy vehicles, jet fuel for the growing fleet of commercial aircraft, 'naphtha' – mixed light fractions – as chemical feedstock, light fuel oil for industrial gas turbines, 'paraffin' for oil heaters, heating oil, and heavier oil for marine engines. At the bottom of the barrel remained a tarry residual oil – 'resid' or 'HFO' for heavy fuel oil – that could be burnt to raise steam in the huge electricity generating stations that began to sprout next door to refineries.

You could divide a barrel of crude differently, according to the current prices for the different fractions. But the product that above all determined the breakdown of the barrel was petrol, or – as it has always been called, confusingly, in North America – 'gasoline' or 'gas'. From 1950 onward the long-standing symbiosis between oil and the car bloomed, with the US to the fore. As the car population expanded so did the road systems. The 'Autobahns' of Hitler's Germany served as a model, and prompted a postwar blitzkrieg on cities and countryside. Throughout the 1950s and 1960s the 'motorways' and 'freeways' unrolled, wide aprons of concrete carrying six, eight or more lanes of motor-vehicle traffic, giving – as the proponents insisted – new freedom to the public, opening fresh horizons and moving the necessities of thriving economies more efficiently. It seemed a good idea at the time.

Meanwhile, breaking down thousands of millions of barrels of crude oil to quench the insatiable thirst of the car made other fractions abundant and cheap, like heating oil and heavy resid. Home-owners installed oil-fired central heating: factories switched from coal to oil; and electricity suppliers began building more and larger oil-fired plant. Imported oil dealt a savage blow to indigenous fuel suppliers; the tide of incoming oil flooded out the coal mines of France and Belgium and gravely imperilled those of Britain and Federal Germany. Coal simply could not compete against the cheap, copious and convenient oil. Slowly but apparently inexorably coal, the fuel that had fired the Industrial Revolution, was finding itself outmoded and discarded. By the mid-1960s officials were planning to run down even Britain's coal industry completely.

As if oil were not enough, its increasingly attractive accompaniment, natural gas, also poured onto the world fuel market. To be sure, you could not readily produce gas in the Middle East and burn it in the Middle West. But the thriving natural gas industry of the US prompted the struggling coal-gas industries of Europe to seek access to the natural gas then used only to light up the night skies around the eastern Mediterranean; the 'flares' from Middle Eastern oilfields were even visible to astronauts. Transnational pipelines soon linked suppliers and users. But long pipelines crossing volatile territory were both costly and politically vulnerable. You could chill natural gas until it condensed into a fiercely cold liquid at −162 degrees Celsius. This so-called 'liquid natural gas' – a linguistic monstrosity, like saying 'solid steam' – usually known as LNG, occupied only one-thousandth the volume of its gaseous equivalent; it could therefore be transported economically for long distances in refrigerated tankers. It could also, however, escape, evaporate and explode catastrophically. On 20 October 1944 an LNG storage tank in Cleveland, Ohio, ruptured; the resulting conflagration burnt out thirty acres of the city, causing widespread death and injury. LNG safety became a long-running issue.

In 1959 drillers in the Netherlands suddenly hit the largest reservoir of natural gas ever discovered, the Slochteren field north of Groningen. By the time Slochteren gas was emerging from cookers in Amsterdam, Belgium, Federal Germany and France, drillers off the east coast of Britain again struck gas: the West Sole field, the first of a whole series of offshore gas fields in the North Sea. To welcome North Sea gas ashore, Britain's gas suppliers embarked on the heroic task of upgrading the country's entire gas-transmission network and gas-fired hardware, both domestic and industrial, to use the new high-pressure high-energy natural gas. Britain's North sea gas conversion programme – initially bitterly controversial, ultimately a stunning success – illustrated vividly how rapidly and dramatically a society could change its energy technology, given the necessary political determination. The lesson may be worth recalling in the 1990s, as we shall see.

Britain's conversion to North Sea gas was only one of the more obtrusive examples of the sweeping changes brought about by cheap oil and gas. Meanwhile, as oil and gas were knocking the bottom out of the price of fuel, the price of electricity was likewise falling. From the late 1940s, electricity suppliers in many countries set off in pursuit of what they saw as the advantages of increasing scale, both of generating plant itself and of the associated transmission and distribution network. Larger generating units converted fuel into electricity more efficiently; larger networks used all the available plant more effectively, reducing the cost of electricity to customers. In 1945, for instance, only one generating unit in Britain – Battersea B at 100 megawatts – could deliver more than 60 megawatts of electricity, enough to light 600,000 100-watt bulbs. By 1965 British electricity suppliers were operating units with an output of 500 megawatts and ordering units of 660 megawatts, a tenfold increase in output per unit. The same scale-up occurred throughout the industrial world. At the same time, transmission networks were extending ever farther, tying even remote corners of Europe and North America together in a vast web.

From the mid-1950s onwards a further factor entered the electrical equation. Electricity suppliers in Britain, the US, and elsewhere found themselves encouraged, subsidized and browbeaten by their governments into involvement with a new technology: nuclear power. The electricity suppliers were dubious. Nuclear power was an unproven and unfamiliar technology, and the suppliers were enjoying the resurgence of a buyers' market in conventional fuels. By the late 1950s coal, oil and gas were all abundant and becoming more so; why should suppliers risk their capital and their planning on a chancy additional technology whose economics were at best unexciting? The governments, however, briefed by their own nuclear advisors, had leverage they did not hesitate to use. In Britain, where the electricity supply industry was by this time 'nationalized', fully owned and operated by a government agency, the industry could be – and was – summarily directed to undertake, in the title of a government White Paper in 1955, 'A Programme of Nuclear Power'. In the US, where the electricity supply industry consisted of many separate companies, some private and some owned by local authorities, the federal government offered generous financial assistance to suppliers willing to build nuclear plants. The federal government also issued a veiled threat that, should suppliers not cooperate, the federal government itself, through its Atomic Energy Commission, would soon be in the electricity supply business. The prospect of federally-funded competition made electricity suppliers blanch; when offered federal funding for cooperative nuclear projects they tended to go quietly.

At length, in December 1963, US General Electric announced that it had sold the Oyster Creek nuclear plant without any federal financing; and many observers, especially from within the reactor supply industry, declared that nuclear power was at last economic. Only much later did it emerge that the Oyster Creek sale and nearly a dozen that followed were subsidized not by the government but by General Electric and Westinghouse themselves, costing the companies hundreds of millions of dollars, in an attempt to stimulate a market for their nuclear hardware. Nuclear economics were to remain swathed in hype and mystery for more than two decades.

Nevertheless, despite the burden of nuclear commitments, electricity suppliers could watch the real price of their product continue to fall, and the sales of electricity and hardware to use it continue to rise. Electricity became the paradigm of modern energy-use. In North America it was 'cheap, clean and convenient', as the advertisements proclaimed; in the Soviet Union, Lenin's dictum declared that 'Communism is Soviet power plus electrification of the whole country'. It mattered not that electricity was not a 'source' of energy but merely an 'energy carrier'. Energy carriers – fuels and electricity – had become so commonplace and ubiquitous that they were taken for granted.

Energy carriers and their end-use hardware had developed a technological symbiosis; each required the other for them both to function. For many millenia the relation between a fuel and its end-use hardware was loose and flexible. You could burn wood or coal interchangeably in the same fireplace or furnace; conversely, you could operate the fireplace or furnace with either wood or coal. Gradually, however, the specifications began to tighten. Without the correct hardware the fuel or electricity was useless; and without the correct fuel or electricity the hardware was likewise useless. In due course this interdependence became highly specialized. You could not drive your car by filling its tank with paraffin or heating oil, much less coal; indeed by the 1950s you could not use just any petrol, but required a grade of petrol appropriate to your particular engine. You could not run a vacuum cleaner on just any electricity; you required electricity of the correct 'voltage' and 'frequency'. Conversely, since the public electricity supply was already in place supplying electricity with predetermined characteristics, you could use only a vacuum cleaner or radio or even light bulb designed to use the available electricity and no other. Differences of system characteristics from place to place added further complications; a US radio, for instance, would not operate from a British wall-socket.

The energy systems of industrial society, consisting of ever more specialized and symbiotic energy-conversion hardware, thus became, if not precisely self-perpetuating, certainly difficult to change in a hurry. Each particular end-use – lighting, cooking, space and water heating, industrial high temperatures, motive power, mobility, materials processing and so on – relied on a progressively more tightly integrated chain of energy conversions. Each link of the chain had to be in place and functioning if you were to attain your final objective. For instance, you might need virgin fuel, at the right time and place in the right quantity. You might need intermediate hardware – say a power station or a refinery – in place and operating. You might need a distribution network – pipeline, cables, railway, tanker lorries or the like – in place and operating. You needed the end-use hardware – light bulb, car engine, electric motor, gas fire, whatever – available and ready to operate. Breaking the chain at any point would deprive you of your final objective. A miners' strike, a power station breakdown, a fallen transmission line, a blown fuse or a burnt-out bulb could all leave you in the dark. Experienced energy users, while enjoying the benefits of modern energy conversion, nevertheless kept a stock of candles handy.

THE BALANCE SHIFTS

Even in the early 1960s a scattering of percipient commentators began to sound warnings about the vulnerability and environmental impact of our energy systems, as we shall describe in Chapter 5. But their monitions were discounted as at best special pleading, at worst pure crankiness. In 1960, however, another development got underway, in far from auspicious circumstances; within fifteen years it was to shatter the traditional foundations of political and economic policy about energy. The major international oil corporations, faced with yet another glut of petroleum, had determined to reduce the price offered to the oil exporters. The governments of the oil-exporting countries decided to band together to resist this corporate pressure; and in September 1960 they formed the Organization of Petroleum Exporting Countries – OPEC. OPEC was created to present a united front against the oil corporations; but the prevailing abundance of oil, including that from domestic wells in the US, and the mutual wariness of OPEC member countries, left the new producers' organization effectively powerless. The oil corporations continued to dictate policy on prices and extraction rates, and discounted OPEC as an empty threat.

Ten years later, in 1970, the position had changed. Worldwide use of petroleum had expanded so rapidly that the producers were hard put to satisfy eager users. In the US, domestic wells could no longer quench the thirst of the home market; the US became a net importer of oil. Against this background OPEC met the oil corporations in Teheran in January 1971, to negotiate new prices and terms. A see-saw fortnight of bitter bargaining produced an agreement to raise the reference price by thirty cents a barrel. In retrospect the outcome hardly seems earth-shaking; but its symbolic implication was unambiguous. The producer countries, not the corporations, were now in charge.

As the months passed the tension mounted. By October 1973 it had reached snapping point. On 6 October Egypt and Syria invaded Israeli-occupied territory. On 8 October OPEC met the oil corporations in Vienna. OPEC asked for six dollars a barrel; the oil corporations flatly refused. Sheikh Zaki Yamani and his Saudi Arabian delegation left Vienna; the conference had collapsed. Within ten days, after a series of ever-angrier confrontations in Washington and elsewhere, OPEC announced not only a unilateral price increase and production cut, but also a complete embargo on shipment of oil to the US and the Netherlands.

OPEC imposed the embargo because the US and the Netherlands were supporting Israel; but it proved a devastating coup to back the unilateral price increase. In the preceding two decades, the body of western industrial society – its buildings, its industrial plant, its transport facilities – had become ever more dependent on a single fuel. Imported oil had become its lifeblood. The OPEC embargo struck the industrial world like a sudden and violent heart attack.

4
ENERGY IN TURMOIL

Oil-users in importing countries reacted to the OPEC embargo and price-rise with blank disbelief. This could not be happening. But it was. The unilateral price increase was alarming enough; but what really shook users was the embargo. What triggered the panic, not only in the US and the Netherlands but in every industrial country, was the dawning realization that in only two decades the oil-users in importing countries had delivered themselves helpless into the hands of their foreign suppliers. The oil-using cars and central heating and power stations would be cold, still and useless without their pungent lifeblood; and nothing the users could do would make any difference until far too late.

In the weeks and months after what became known as the 'oil shock', oil was transformed from a commodity dealt with by commerce, into a policy issue dealt with by governments. Nor was this transformation limited to oil alone. In the US, the onset of the winter of 1973–4 demonstrated that, even before the OPEC embargo, supplies of heating oil and natural gas would not suffice to meet the peak demand. Officials, however, blamed OPEC, as schools and factories were closed and their shivering cohorts were sent home, sometimes to homes not much warmer. By mid-November 1973 gasoline rationing was threatened, giving a boost to a project long stymied in the courts. British Petroleum and Standard Oil of Ohio (Sohio) at last won Congressional approval for an oil pipeline through the Alaska wilderness from Prudhoe Bay on the North Slope to Valdez on the south coast of Alaska.

The British government, too, hastened to approve proposals to speed up production of its indigenous oil, from the North Sea. The Conservative government of Edward Heath even attempted to 'national-ize' coastal land for sites to build offshore platforms, a startling demonstration of how thirst for fuel could override the historic principles of an entire political movement. This controversy, however, was overtaken by yet another, over yet another fuel.

In late 1973, Conservative government policy dictated rejection of a wage claim from British coal-miners. They began a 'work to rule' that so reduced output from the mines that stocks of coal at power stations fell alarmingly low. To spin out the remaining coal, in January 1974 the government decreed that industry must work only a three-day week, and suspend production the rest of the time. A government publicity campaign hammered the slogan 'SWITCH OFF SOMETHING – NOW!'. One hapless Conservative Minister proposed that people save electricity by brushing their teeth in the dark. By February 1974 the miners had called an all-out strike. Prime Minister Heath retaliated by calling a snap election to decide 'who rules the country – the government or the miners'. The voters decided that it would not, at any rate, be the Heath government, which lost the election. The confrontation left lasting scars on Britain's body politic; it also underlined the crucial role that energy carriers had assumed in public policy.

Some had long seen them in such terms; but it took the winter of 1973–4 to drive the point home to politicians. The convergence of problems with all the major energy carriers – oil, coal, natural gas and electricity – also demonstrated the scope and complexity of the subject. Ironically, governments came to recognize their responsibilities most acutely by finding that they could not in fact do anything – at least not anything easy. In the winter of 1973–4 the powers-that-be in western industrial society found out about the power that wasn't: their presumed power to ensure continuing conversion of energy under human control to keep society functioning. For the individual energy-user, however, the lessons were much more immediate and much less metaphysical. A modern 1960s house with oil-fired central heating can be very cold in January when the fuel-tank gauge reads 'empty'. 'Cooking with gas' may mean peanut butter sandwiches if the mains pressure drops. Where was Moses when the lights went out? In the dark – and citizens throughout industrial society knew how he felt. You can be a two-car family and still have to walk; and in a city laid out for the internal combustion engine, getting there on foot is much less than half the fun. Many a fuel-addict had to kick the habit cold turkey; and the withdrawal symptoms were acutely discomfiting.

Even when your energy carriers were not cut off, your government tried to make you feel guilty about using them. First came exhortation: switch off something, leave your car in the garage, watch less TV. Britain, for instance, began 1975 under an onslaught of government advertisements featuring the brusque rubber-stamp directive 'SAVE IT!'. Governments also adopted more 'governmental' measures, reinforcing exhortation with legislation. In Britain the 'SAVE IT' campaign was accompanied by Fuel and Electricity Orders imposing restrictions on the use of these energy carriers. The restrictions were more cosmetic than draconian. They banned electric lighting of advertising signs in the daytime; switched off some motorway lights; and decreed lower speed limits and maximum permissible indoor temperatures. Other countries imposed similar measures; but whether they made much practical difference to the use of energy carriers is debatable. The official restrictions were often of a kind that is, in a democracy, impossible to enforce. How do you ensure that some office manager, defying the 'general good', unwilling to risk upsetting further a staff already jangled by power cuts and public transport jams, has not left a thermostat above the legal limit? Breathless commentators warned that the authorities would have to create Energy Police, with extraordinary powers of entry and summary arrest. To be sure, some constraints were comparatively easy to enforce. In the Netherlands the government made driving a car illegal on a sequence of Sundays. Anyone breaking that particular law stood out on the quiet street like a streaker at a prayer meeting.

Such measures often appeared to be devised primarily to inflict an official inconvenience on the public, to impress it with the severity of the energy problem and the determination of the authorities to tackle it. In Britain and several other countries, governments ordered an early shutdown of TV broadcasting. As a way to restrain energy-use the measure was probably counterproductive, being just as likely to prompt people to turn on bedroom fires earlier than usual. But it irritated the maximum number of people, and could be centrally enforced; it was therefore a comparatively easy way for governments to act tough about energy use.

For all the government exhortation and heavy-handedness, however, the most effective restraint on using energy carriers from 1974 onward was the precipitous economic recession into which the world toppled, with the sonorous inevitability of a felled redwood. The OPEC price-rise was by no means the only cause; but it grievously aggravated economic and social tensions, and drastically undermined an already teetering confidence in the existing order.

As the recession intensified, firms unable to withstand the steep increase in fuel and electricity prices went to the wall. Others survived, but could not find enough buyers for their products and services, and cut back sharply on hiring and staff. Healthier firms battened down the hatches and mothballed expansion plans. Those with jobs often lost them. The young left schools and colleges and found themselves on the dole. All told, it was undoubtedly an effective way to reduce the use of fuels and electricity; but it was not quite what governments had in mind. Even so, it was a striking manifestation of a phenomenon that governments would subsequently endorse: 'energy conservation' through the price mechanism, with a vengeance. If you do not have a job you do not have much money. If you do not have enough money you cannot afford so much fuel and electricity. If you cannot afford fuel and electricity you do not use it; or – for instance in the case of gas or electricity from a public supply – you use it, cannot pay the bill, and have your supply disconnected. It is certainly a tidy and enforceable way to reduce the use of energy carriers, and would probably appeal to governments, if it did not have such unsettling side-effects. Governments could not be easy in their minds about an 'energy conservation' policy that entailed freezing the electorate to death. Nevertheless, the idea of 'energy conservation' as a form of self-denial, voluntary or imposed, ascetic, romantic or willy-nilly, took firm hold on the public imagination, usually associated with other economic discomfort.

By this time a new linguistic usage had taken root. In the initial weeks after the OPEC embargo, political and media commentators still referred to the 'oil crisis', or more broadly to the 'fuel crisis'. By early 1974, however, as problems with natural gas, coal and electricity piled up, the 'crisis' had become, by almost unanimous agreement, the 'energy crisis'. This form of speech was to have a lasting influence on public perception of the issues. One of the first and most persistent effects of the usage was to blur the distinctions between the different fuels and electricity – to encourage the impression that they were all 'energy sources' or 'energy supplies', and more or less interchangeable. This impression in turn underpinned the drive to find 'substitutes' for oil. The urge to reduce OPEC's leverage and indeed to cock a snook, if possible, at the upstart ingrate Arabs, became inextricably entangled with the purported necessity to find a 'substitute' for oil in the twenty-first century, when the wellheads had ceased to pour forth their largesse once and for all. The consequent confusion of objectives, constraints and time-scales further muddied policy waters already all too murky.

Governments embarked on both short and long term programmes to alleviate the 'energy crisis'. Their efforts were not markedly successful, as we shall discuss in the next chapter. They were also punctuated with further complications from the news headlines. On 17 March 1978 the supertanker Amoco Cadiz broke up in the Bay of Biscay and discharged over 200,000 tonnes of crude oil into the coastal waters of Normandy. The wreck of the Amoco Cadiz was only the largest and most dramatic of a succession of major tanker accidents fouling the world's oceans and coastlines. Nor did oil have to pass through a tanker to reach the ocean. In the summer of 1979 the first major production well drilled off the coast of Mexico, Ixtoc 1, ran wild; it proved impossible to cap for nearly a year. The oil that poured from Ixtoc 1 defaced shorelines as far north as Louisiana in the US. Nor was the negative impact of oil-production limited to pollution. On 27 March 1980, in the Norwegian sector of the North Sea, the semisubmersible rig Alexander Kjelland capsized, costing 123 lives.

On 28 March 1979 the second unit at the Three Mile Island nuclear power station near Harrisburg, Pennsylvania suffered a nightmarish accident. A leaking valve allowed cooling water to escape, uncovering the nuclear fuel; the fuel melted, releasing potentially lethal radioactivity.

Panic and confusion ran wild, as officials contradicted each other and thousands of families fled the area. Although the operators finally brought the reactor under control and averted a catastrophic release of radioactivity, the accident cast a pall over the already uncertain future of nuclear power in the US, and produced reverberations worldwide. In Federal Germany, in response to public opposition, the premier of Lower Saxony refused permission for construction of what would have been the world's largest civil nuclear installation, near the village of Gorleben. In Sweden the government agreed to a referendum on the country's nuclear programme; a year later Swedish voters elected to terminate the programme and close down all the reactors by 2010. Many other countries looked long and thoughtfully at their previously ambitious nuclear plans.

CRISIS AGAIN

Meanwhile, in January 1979, the world oil scene had once again been jolted to its deepest strata. The Middle East had erupted again, uglier than ever. The Peacock Throne of the Shah of Iran, which had looked progressively shakier throughout the autumn of 1978, was finally and messily overrun. Within days the evanescent glut of crude turned into a shortfall, as Iranian output disappeared. No sooner had it begun to flow again than the Soviet Union, in December 1979, invaded Iran's next-door neighbour, Afghanistan; many viewed the invasion as a grave military threat to the west's precious oil supplies from the Middle East. The US and the Soviet Union squared off, bristling with hostility and armaments. The new decade of the 1980s opened in an atmosphere like that of the Cold War 30 years earlier, with access to oil a key factor.

In January 1980, according to the OECD, the average price of world oil was about US$26 a barrel. Compared to the pre-1973 price of less than US$3 a barrel the rise seemed to be nearly tenfold; but the comparison was misleading. Oil was so important in modern society that its 1973 price-increase had triggered a worldwide surge of inflation. Prices of everything everywhere rocketed through the 1970s, often at more than 10 per cent per year; by 1980 a nominal dollar would buy much less than it would have in 1973. Economists compensate for inflation by comparing so-called 'real prices' – that is, prices adjusted so that the money in which they are expressed will buy the same goods or services in different years; this cancels out the effect of inflation so that the prices compare like with like. Even in real prices, however, oil at $26 a barrel in 1980 was substantially more expensive than oil at $3 a barrel in 1973.

Moreover, by 1980 different OPEC member countries were asking very different prices for similar oil; OPEC's previously united front was showing signs of serious strain. Saudi Arabia, with enormous reserves, was content to produce generous quantities of oil and sell it a price of some US$24 a barrel. Iran and Libya, by contrast, wanted to cut back production and push prices much higher, to more than US$40 a barrel. Their stated aim was to charge a price almost – but not quite – as high as the cost of novel alternatives like synthetic oil or 'synfuel' made from coal. Saudi Arabia feared that this policy would cause such economic disruption in importing countries that oil-use would collapse entirely. The tension between the OPEC factions took an ugly turn later in 1980 when Iraq and Iran clashed in a grim and futile conflict that was to drag on through eight grisly years. The Gulf, with the combatant countries to either side, became a war-zone; the vast tankers laden with oil for users around the world had to run a gauntlet that grew ever more hostile.

Nevertheless, by early 1981, despite the various threats to oil supplies, commentators were noting that the world seemed to be 'awash with oil', and prices began edging down again. Since 1973 the combined effects of economic recession, users switching to other energy carriers, and improving efficiency had brought about a steady decline in the use of oil in industrial countries. Natural gas, once regarded as not worth the cost of gathering and delivering, had emerged as a major worldwide energy carrier in its own right; international gas-pipelines and LNG facilities proliferated. One of the more startling proposals was for a pipeline extending from gasfields deep in Siberia, westward across the one-time 'Iron Curtain' and into western Europe. Some western observers looked with skepticism – or indeed outright paranoia – at the prospect of western Europe coming to depend for gas supplies on what US President Ronald Reagan called the 'evil empire' to the east. But Soviet gas reserves were impressive; the Soviet Union was ravenous for hard currency; and the terms it was offering, both to western construction firms and to western gas suppliers, were too tempting to ignore. Besides, said the proponents, one way to undermine putative Soviet aggressive intent would be to make the Soviet Union dependent on income from customers in western Europe. From having been something of a poor relation of oil, natural gas was turning into oil's most challenging competitor.

The immediate aftermath of the 1973 price rise evoked in some quarters headlong enthusiasm for nuclear power 'as a substitute for oil'. France in particular decreed in 1974 that state-owned Electricité de France would erect a cavalcade of nuclear plants; and so it did, at a breakneck pace, six or eight large units ordered each year through the 1970s. Elsewhere, however, under the influence of escalating costs and construction times, erratic performance, and concern about safety and waste disposal, the spasm of nuclear enthusiasm abated almost as rapidly as it had set in. Nuclear manufacturers in industrial countries turned their gaze toward the developing countries, hoping to keep their order books active by exports.

On 7 June 1981, however, the international nuclear business received a nasty shock. Israeli planes launched a raid on Iraq's Tammuz research reactor near Baghdad, destroying it with bombs and rockets. Israeli Prime Minister Menachem Begin declared that if the reactor had been allowed to start up it would have provided materials for Iraq to manufacture nuclear weapons. Iraq protested that the reactor had been inspected by the United Nations International Atomic Energy Agency (IAEA) only a few months earlier, and given a clean bill of health; but Israel was adamant. The episode further increased Middle Eastern tensions; it also did little for the credibility of the so-called 'safeguards' system operated by the IAEA, intended to ensure that civil nuclear activities did not open the way to nuclear weapons. Suddenly it appeared that the only countries eager to import new nuclear plants were countries whose motives for acquiring nuclear technology were at best ambiguous, among them Pakistan, Argentina and Libya.

Be that as it might, what really stymied the nuclear salesmen was not the potential weapons-implications of the technology but its cost. Most prospective Third World purchasers simply could not afford it. Furthermore the World Bank, otherwise a key source of capital for major energy developments, stubbornly refused for purely economic reasons to back any nuclear project whatever. The international market for nuclear power plants withered almost as quickly as the home markets.

TROUBLE IN THE AIR

For nearly two decades the original industrial fuel – coal – had been overshadowed by oil, gas and nuclear power. By the early 1980s, however, coal was again emerging from the shadows. Major oil corporations like Shell, British Petroleum and Exxon bought up coal reserves and coal companies in the US and elsewhere, and declared that world coal-use would double or even triple by the turn of the century. In 1980 coal-users were burning about 2.6 billion tonnes a year; but only about 200 million tonnes of this was traded internationally, of which 150 million were for making steel and only 50 million for power stations and industrial boilers. Analysts foresaw a dramatic increase in international trade in coal, with vast bulk carriers crisscrossing the oceans and enormous coastal terminals for loading and unloading the cumbersome cargoes.

However, even as coal-producers saw their prospects brightening, another shadow fell over them – an invisible, insidious shadow that lengthened relentlessly. Of course some noxious side-effects of burning coal had been all too evident for centuries. Disasters like the 'killer smog' that claimed thousands of victims in London in December 1952 had at last compelled some governments to demand stringent controls on the smoke emitted from coal-burning, like Britain's Clean Air Act of

1956. This Act created 'smokeless zones' in which coal could not be burned in an open fireplace or any other hardware that allowed visible smoke to come out the chimney.

The British Act and similar legislation elsewhere gradually reduced the emission of visible smoke from coal-burning. It also helped to encourage a switch from domestic fireplaces to electric fires; in effect, the energy in the coal was delivered to the household not in burlap bags but by wire, from the power station. Unlike a domestic fireplace, a coal-burning power station could include complicated and expensive arrangements for controlling the byproducts of combustion. In Britain, for instance, the Clean Air Act advocated that smoke from power-station boilers be fed into the bottom of enormous chimneys or 'stacks', sometimes more than 200 metres tall. The hot smoke emerged from such a stack fast enough to rise high into the atmosphere. The power station engineers considered that this arrangement would 'dilute and disperse' the smoke; by the time it returned to earth it would no longer be a problem.

Unfortunately, however, you cannot clear the air simply by eliminating the plume of submicroscopic particles you see as 'smoke'. Coal contains not only carbon and hydrogen but also variable proportions of sulphur and nitrogen. When you burn coal you burn not only the carbon but also the sulphur and nitrogen; and you get not only carbon dioxide but also a mixture of sulphur oxides and nitrogen oxides – what chemists call 'SOx' and 'NOx': invisible, pungent, choking gases that are toxic to breathe, even in low concentrations. If you bubble carbon dioxide through water you get carbonated water, a weak acid that makes a refreshing beverage. However, if you bubble SOx and NOx through water you get strong acids, including sulphuric and nitric acid. SOx and NOx from burning coal react this way with moisture in the air. When the moisture condenses, it falls to earth as 'acid rain'.

The expression was coined in 1872 by Robert Angus Smith, Britain's first 'alkali inspector', appointed by the government to monitor pollution from factories. However, despite his efforts and those of his successors the air over Britain's cities remained smoky and acidic until the early 1960s, after the smokeless zones and tall stacks of the Clean Air Act took effect. The tall stacks dramatically reduced the local concentrations of smoke particles and gases; but they worked if anything too well. By the time the SOx and NOx from Britain's power stations returned to earth, a substantial fraction were settling not on Britain but on Scandinavia; and the Scandinavians began to object.

The issue was already poisoning the air metaphorically as well as literally by June 1972; at the United Nations Conference on the Human Environment in Stockholm, the Swedish hosts emphatically recorded their displeasure at their uninvited imports. They found their lakes and rivers growing so acidic that fish could not survive; but their protests were unavailing. Nor was Britain the only country to export its undesirable combustion gases. Tall stacks in the northern US wafted their burden of SOx and NOx across the border into Canada; and Federal Germany, East Germany, Czechoslovakia and Poland swapped air pollutants according to the prevailing winds. By the early 1980s acid rain had become a major political and diplomatic controversy.

The 'importing' countries, including Norway, Sweden, and Canada, accused the 'exporters' – especially Britain and the US – of poisoning waterways, killing fish and other aquatic animals, destroying forests and even defacing historic buildings. But the British and US governments insisted that the evidence linking acid emissions and ecological damage was inadequate and scientifically suspect, and demanded further research before spending the staggering sums necessary to mitigate the effects of acid rain.

Coal, to be sure, was not the only source of atmospheric SOx and NOx. Fuel oil, too, contained sulphur; and the main producer of NOx was not power stations at all, but internal combustion engines in cars. Nevertheless the acid rain issue confronted current and prospective coal-users with yet another dilemma. They could equip coal-fired power stations with technology to remove the SOx from the smoke before discharging it to the atmosphere; but the various designs of 'flue-gas desulphurization' or FGD plant all shared several unattractive features. An FGD plant for a large modern power station might cost more than £200 million; it would require enormous quantities of limestone or other sulphur-trapping materials; and it would reduce the output of electricity from the plant. Technical innovations promised to provide more attractive options in the 1990s, as we shall describe in Chapter 8; but in the meantime, burning coal would become either more environmentally unacceptable or more expensive – if not indeed both.

In 1985 Dr Joe Farman of the British Antarctic Survey found that the layer of ozone in the atmosphere high above Antarctica was measurably less dense than hitherto. The implication of this apparently exotic and obscure scientific discovery was, paradoxically, immediate and hair-raising. Ozone in the upper atmosphere helps to screen out the biologically dangerous ultraviolet radiation reaching the earth from the sun. If the ozone layer is thinner, more ultraviolet reaches the surface of the earth, increasing the incidence of skin cancers and eye cataracts in people, and disrupting the delicate balance of the earth's biological systems. Not only the rain but also the sunshine was becoming positively unhealthy.

The thinning of the ozone layer was rapidly confirmed by other scientists, who had known the likely cause since the early 1970s. Unlike acid rain, ozone destruction is not a consequence of human energy use; it is triggered by a group of artificial chemicals called 'chlorofluoro-carbons' or CFCs, hitherto believed to be utterly benign and employed for precisely this reason. Ironically, however, the chemical inertness and stability of CFCs, that has made them so useful as spray-can propellants and blowing agents for plastic foam packaging, also allows them to drift upward through the atmosphere unaltered, until they reach the stratosphere, and the layer of ozone it includes. At this height the intense radiation from the sun, bombarding the CFC molecules, may fracture them, freeing chlorine. The chlorine then reacts vigorously with any nearby ozone, changing its chemical nature and making it much less effective as an ultraviolet shield.

The 'hole' in the ozone layer drove home a startling realization: you could threaten the fabric of life on earth by using something as obviously innocuous as a spray deodorant. Although the ozone issue did not arise directly from using energy, it made many people realize for the first time that even the most apparently trivial or casual human activities – including using energy – could have genuinely global implications. So could preparing energy for use.

CHERNOBYL, THE GREENHOUSE AND AFTER

On 26 April 1986, at 1:23 a.m. local time, unit 4 of the Chernobyl nuclear power station in the southwestern Soviet Union erupted in a catastrophic explosion. The world beyond Soviet borders, however, was unaware of the disaster until two days later, when instruments at the Forsmark nuclear power station in Sweden, about a thousand kilometres from Chernobyl, signalled abnormal levels of radioactivity. The Forsmark plant was evacuated; but Swedish officals rapidly ascertained that the radioactivity came not from their own reactors but from deep in the Soviet Union.

In the days that followed, panic and confusion ran wild over much of the northern hemisphere, as the cloud of radioactivity from Chernobyl rode the winds across the face of Europe. Fragmentary information from Soviet sources was compounded by a deluge of contradictory pronouncements by Western governments, nuclear agencies and news media. Some governments warned their citizens to keep children and animals indoors, and imposed bans on eating or moving fresh produce. Others, notably the governments of Britain and France, declared that Chernobyl posed no threat to public health – only to be compelled to acknowledge, weeks later, that radioactive fallout had contaminated parts of their countries seriously enough to require stringent controls on agriculture.

Sheep farmers in north Wales and northwestern England, for instance, were suddenly faced with official bans on slaughtering or marketing their lambs; and despite repeated official assurances, the controls remained in force in some areas for years. The impact of the Chernobyl explosion even contaminated Lapp reindeer in the far north of Scandinavia. Its effects closer to the site became clear only four months later, when a top-level Soviet delegation reported its findings to a conference organized by the International Atomic Energy Agency in Vienna. In addition to 31 direct fatalities, the accident had sent nearly 300 more victims to hospital with acute radiation sickness, and necessitated immediate evacuation of some 135,000 people from the surrounding area. In the months and years that followed, the extent and persistence of the fallout from the explosion far exceeded initial expectations. Soviet authorities estimated the early cost of the accident at some eight billion roubles – about eight billion pounds sterling; but they faced further enormous bills for belated evacuations and clean-up even after the beginning of the 1990s.

The Chernobyl accident, with its unprecedented financial conse-quences and its devastating social impact, both short and long term, profoundly deepened the pall that had been descending over nuclear power in the 1980s. Even those countries that had been proposing to expand their nuclear programmes abruptly put them on hold. Their turn away from nuclear power was accelerated by a precipitous fall in the world price of oil. Precarious through the early 1980s, by 1986 the OPEC cartel appeared to be losing its grip entirely, as member countries exceeded their agreed production quotas and oil poured onto a world market unable to soak it up. A series of emergency meetings failed to reassert OPEC control over its recalcitrant members; at one point the base price of oil threatened to fall below US$10 a barrel.

In the 1970s, western oil-importing countries would have rubbed their hands in glee at the sight of OPEC in disarray. By the late 1980s, however, a collapse of oil prices would have been as unwelcome in Washington as in Riyadh, for two main reasons. One reason was already long established. Both oil-suppliers and oil users cared less about the exact level of the oil-price than they did about a stable, predictable price, that did not vary too much. Wild variation in the oil-price left planners unable to assess the economic status of proposed investments in energy hardware – everything from central heating to power stations – because they could not predict what the equipment would cost to run throughout its working life. In the aftermath of the oil-price rises of the 1970s, industrial countries had invested vast sums in energy hardware that would be economic when oil cost more than US$20 a barrel, but seriously uneconomic if it cost less than half that much.

The second reason for concern about falling prices of oil and other fossil fuels was both unexpected and potentially far more grave. For some two decades a handful of scientists had been watching with growing concern the buildup of the gas carbon dioxide in the earth's atmosphere. As described in Chapter 2, carbon dioxide in the atmosphere, like the glass of a greenhouse, traps slightly more sunlight energy inside it, raising the temperature at the surface of the earth. This so-called 'greenhouse effect' is essential to the existence of life on earth. Without an atmosphere to retain sunlight energy, the earth would be as uninhabitable as the moon. However, even an apparently insignificant shift in the delicate energy balance within the global 'greenhouse' might have drastic consequences.

As yet we cannot adequately comprehend the complex energy-patterns that drive the circulation of the atmosphere and the oceans. But six of the hottest years ever recorded occurred in the 1980s; and many scientists are now convinced that the 'greenhouse effect' is already intensifying. An impressive convocation of more than 300 experts from 48 countries conferred in Toronto, 27–30 June 1988, on 'The Changing Atmosphere: Implications for Global Security'. The opening sentence of their concluding statement was uncompromising: 'Humanity is conducting an uncontrolled, globally pervasive experiment whose ultimate consequences could be second only to a global nuclear war.' They called for immediate policy measures to mitigate its impact. Nothing could be done about the carbon dioxide and other 'greenhouse gases' – including CFCs, methane and nitrous oxide – already added to the atmosphere; their effects would continue to accumulate inexorably. At best the world's governments could act to reduce further discharges of greenhouse gases. CFCs, already implicated in damage to the ozone layer, were being phased out; but the phase-out should be accelerated. Methane, from leaking natural-gas supplies, rice paddies and – bizarrely enough – flatulent cattle, would be harder to suppress; so would nitrous oxide, much of which came from agricultural fertilizers. Carbon dioxide, however, accounted for an estimated 50 per cent of the overall effect; and most of it reached the atmosphere as a result of burning fossil fuels. The Toronto conference called on the countries of the world to reduce their carbon dioxide emissions by 20 per cent by the year 2005, with further reduction thereafter.

That would mean, above all, making dramatic cuts in the amount of fossil-fuel energy that people use. Commentators warned that aiming for such a target, much less reaching it, would demand political will on a scale never hitherto manifest even in total war. Nevertheless, even senior politicians like British Prime Minister Margaret Thatcher, US President George Bush and Soviet leader Mikhail Gorbachev proclaimed the urgent gravity of the issue. Entering the 1990s, the greenhouse effect promised to reshape human energy-use in a wholly new direction: but which direction?

While the politicians grappled with panoramic global strategies, they were jolted by a barrage of reminders that using fuel poses not only long-term but immediate problems. On 1 June 1988 an underground explosion at the Stolzenbach mine in Federal Germany took the lives of 57 coal-miners. On 6 July 1988, the Piper Alpha oil-production platform in the North Sea was shattered by an explosion that killed 167 men. On 4 June 1989 a cloud of gas from a leaking pipeline in the Soviet Union was ignited by a spark as two trains passed nearby; the blast wrecked both trains, killing over 400 people and hospitalizing hundreds more. On 24 March 1989, the oil tanker Exxon Valdez wandered from the marked channel south of the Valdez oil terminal in Alaska, ran aground and tore open its hull, disgorging 11 million gallons of crude into the rich coastal waters, polluting hundreds of kilometres of pristine wilderness coastline in the worst oil-pollution disaster to strike the US.

In the 1990s, fuels and electricity are so cheap that people are using ever more of them. But the long-term implications of acid rain, toxic wastes, and the greenhouse effect, underlined by immediate, ugly and expensive mishaps involving both fossil-fuel and nuclear technology, indicate that humanity's voracious appetite for fuels and electricity is getting out of hand. Unless we do something about it, we may soon make our own planet uninhabitable.

5
SUPPLYING DEMAND, DEMANDING SUPPLY

As the flames subside and the welcome warmth begins to fade, you glance around you at the encroaching darkness and shiver slightly. Only one gnarled twig remains from the pile you collected before sundown. As the damp chill of the night creeps into your bones, you determine that tomorrow you will not give up gathering firewood quite so quickly . . .

Humans are always thinking about the future, and trying to act accordingly. Even at such a basic level as a campfire, using fuel requires foresight: how much fuel will you need, over what timescale? When and how will you get it? If the responsibility is yours alone, you yourself make the decisions and do your best to carry them out. As we shall discuss in Chapter 9, most people in the world still use fuel-energy at this basic level – if indeed they can. Over the centuries, however, as described in Chapter 3, industrial societies have progressively subdivided the responsibilities involved in using energy.

Well before the time of John Londoner and his contemporaries, specialist suppliers gathered, transported, bought and sold wood and coal for fuel. To succeed in business they had to anticipate how much they could sell to whom at what price. They could obtain the hardware they used – saws and axes, picks and shovels, horses and carts – with little difficulty or expense, augment it quickly if the market expanded, or set it aside if the market contracted. If, however, a fuel supplier considered investing in much more expensive hardware, like a steam-driven mine-pump or sawmill, he would want to know beforehand how much of its output he could sell at what price, over a period long enough to repay his investment and return him a profit. He had to 'forecast' how his customers would behave in the future.

PLANNING SUPPLY

As public gas suppliers and – about a century later – electricity suppliers came on the scene, they had to invest in much more complicated and costly hardware. Moreover, they themselves had to plan to buy the coal they would need to produce their gas and electricity. Their forecasting had to be accordingly much more elaborate, anticipating the future behaviour not only of their customers but also of their coal-suppliers. Customers bought wood and coal in individual batches, at an agreed price per batch. On the other hand they got gas and electricity, so to speak, 'on tap'; they opened gas jets or threw switches as and when they wanted the gas or electricity, and expected each to be available at that instant. Suppliers had to indicate beforehand how much each unit of gas or electricity used would cost; they had to measure how much a customer used, and bill the customer later.

Gas suppliers could store a volume of gas in 'gasometers', for customers to draw on when they wished. Electricity, however, could not be stored. Electricity suppliers therefore had to plan for enough generating capacity to supply the maximum amount of electricity their customers all together might want to use at a given instant – the so-called 'peak demand' for electricity. They had to forecast not only how much electricity customers might want, but when. The terminology they employed indicated the attitude they adopted. They had to plan to meet a 'demand' for electricity that was distinctively different from the 'demand' that a trader might encounter in the sort of marketplace described in books on economic theory. In a 'free market', traders compete to supply a commodity to customers, adjusting their prices until they are all selling at a price just high enough to persuade just enough customers to buy just enough of the commodity so that the available supply just fulfils the customers' demand – 'clearing the market' of the commodity, the suppliers and the customers. As described in Chapter 3, however, public supplies of gas and electricity are 'natural monopolies': a customer is tied to whichever supplier has installed the gas-main or the electricity-main. Unless otherwise restrained, the monopoly supplier of gas or electricity can charge any price whatever; the customers's only recourse is to refuse to use the supply.

As gas, and subsequently electricity, became increasingly essential energy carriers in industrial society, governments intervened. In exchange for granting a 'franchise' to a supplier in a particular locality, a government would impose laws and regulations that the supplier had to obey. The government controlled the prices the supplier could charge, and often stipulated an obligation to supply any customer in the locality with however much gas or electricity the customer wanted, when the customer wanted it. From this perspective, a customer's 'electricity demand', for instance, was effectively non-negotiable. If a customer demanded electricity at a given instant by turning on a switch, the supplier had to deliver the requisite electricity instantly, or face official retribution.

This responsibility became fundamental to the outlook of gas and electricity suppliers, and coloured their approach to forecasting and planning. Under their influence, the consequent assumptions in due course carried over to 'energy planning' in general. As the supply organizations grew larger and their customers more numerous and varied, forecasting and planning grew apace. Oil and coal companies likewise found that they had to look ever farther forward. They were caught up in a rush of industrialization. Customers were hungry for their products, especially petrol; they did not need to stimulate demand, but were hard-pressed to keep up with it. Forecasting could be done by thinking of a number and doubling it.

FORECASTING 'DEMAND'

The second world war changed everything. Aerial bombardment of cities and industrial installations left Europe devastated, its fuel and electricity facilities crippled, its industries in ruins. The aftermath of the war found governments confronting the need to rebuild their societies, to allocate scarce resources, skills and finances to where they could do the most good. In Britain, for instance, the onset of war led the government to create in 1942 a separate Ministry of Fuel and Power, charged with responsibility for this vital sector; and its responsibilities remained onerous long after V-E – 'Victory in Europe' – Day. From the second world war onward, governments assumed an ever more prominent role in forecasting and planning for energy.

A forecaster, almost invariably working either for the government or one of the fuel or electricity suppliers, would begin by collecting suitable data. How much coal, oil, electricity and other energy carriers had domestic sources produced in the preceding year? How much had been imported, exported or added to stocks? How much had been used, and by whom – households, commerce, industry, transport? How did these quantities compare with those for preceding years? Plotting the quantities year by year on a graph allowed a forecaster to identify a 'trend', and estimate how the graph might continue in following years – 'extrapolate the trend'. After 1950, in western Europe, emerging at last from the ravages of the war, all the graphs sloped steeply upward to the right. New housing, new industrial plants, new cars, all contributed to the sharply increasing demand for fuels and electricity.

A government, of course, gathered information not only about fuel and electricity use but about all the financial transactions within and across its borders – so-called 'statistics', meaning 'information useful to the state'. Analyzing these statistics, the government could deduce a composite figure for the entire economic activity of the country: its so-called 'Gross National Product', or GNP. Assuming that more and larger financial transactions meant improved living standards for the population, governments aimed to increase their GNP year by year, achieving what they called 'economic growth'.

Fuel and electricity forecasters noticed what they considered a striking correlation: a country's percentage increase in GNP mirrored closely the percentage increase in using fuels and electricity. If economic activity increased by 5 per cent from one year to the next, so did the use of fuels and electricity. To reduce all economic activity to a single number, and all energy-use to another single number, meant aggregating together many widely different quantities, devising equivalents and taking averages; but it greatly simplified forecasting and planning – especially when the plans had to be spelled out concisely to political leaders or shareholders to enlist their support.

Fuel and electricity suppliers drew up forecasts to guide their planned investment in supply facilities. The scale of these facilities – mines, power stations, oilfields, refineries – was growing ever larger; a single refinery or power station might take several years to design and construct. Before embarking on such an expensive project, a supplier wanted to be sure that customers would be out there waiting, ready to buy its output when it eventually came into operation. As the time-scale for establishing new supply facilities got ever longer, forecasters had to raise their sights higher and higher.

NUCLEAR UPS AND DOWNS

None raised them so high as the promoters of the new wonder-technology, nuclear power. Throughout the 1950s the expanding economies of western industrial countries devoured as much fuel-energy as they could obtain. In Britain, for instance, government planners in the Ministry of Fuel and Power feared that even the rapid build-up of domestic coal production could not keep pace with the country's voracious appetite for fuel. In February 1955 the British government issued a White Paper describing 'A Programme of Nuclear Power', to augment the available coal by burning uranium to generate electricity. In 1956, Egypt's Prime Minister Nasser abruptly nationalized the Suez Canal, threatening oil transport from the Middle East and triggering an abortive military response from Britain and Israel. British planners forthwith tripled the size of their incipient nuclear programme. The planners acknowledged that the electricity from the proposed nuclear power stations would be more expensive than that from coal-fired stations; but they were confident that nuclear power costs would come down, and in any case their forecasts showed that Britain might not produce enough coal nor be able to import enough oil to meet demand.

They were wrong on all counts. The nuclear plants took longer to build and cost more than the 1955 programme foretold. Mechanization of British mines boosted coal production to new heights; and cheap oil from the Middle East flooded into Europe. By 1960 British planners were compelled to scale down and stretch out the nuclear power programme; and even so the chairman of the state-owned Central Electricity Generating Board, Christopher Hinton, himself a brilliant nuclear engineer, continued to consider the programme premature, too large and too expensive.

Throughout the 1950s and into the early 1960s, British planners persistently underestimated the rate of increase of the country's electricity demand. The winter of 1962–3 struck with a severity that overwhelmed Britain's electricity system; blackouts and power failures left Britons freezing in the dark. The political repercussions were immediate and sweeping. British electricity planners rewrote their forecasts with much larger numbers, and embarked on a massive expansion of the country's electricity supplies, ordering an array of new power stations larger than any previously built – some to burn coal, and some to burn heavy residual oil from refineries. After a furious controversy they also launched a second generation of nuclear power stations, the so-called 'advanced gas-cooled reactors' or AGRs.

Before the end of the decade, however, both their forecasts and their plans were looking badly awry. The enormous power stations were not merely months but years behind schedule, so much so that in 1968–9 the British government held a formal inquiry into the delays. However, having underestimated the growth rate of electricity demand for a decade, the forecasters had overcompensated: the anticipated steep increase did not after all arise. Accordingly, the delayed stations were not missed; but they still had to be paid for, even though they generated neither electricity nor income until long after their intended start-ups. Moreover, even against this generally unimpressive background, the first of the second-generation AGR nuclear stations, Dungeness B, was in a class by itself. Ordered in 1965 for service in 1970, it did not start up its first reactor until 1982; and by the 1990s, a quarter-century after its inception, it was still not in full commercial operation. The British AGR programme was an early demonstration that forecasting and planning for nuclear power could reach extremes of wild inaccuracy far beyond those encountered in other energy forecasting. Nuclear programmes elsewhere soon corroborated this.

In the US, for example, as described in Chapter 3, nuclear power plants ordered in the 1960s were mostly subsidized by Westinghouse and General Electric in hopes of creating a market. In 1973, utilities ordered 41 nuclear plants, more than in any previous year. The nuclear industry seemed at last to have made its long-awaited breakthrough to commercial credibility; and – according to nuclear proponents – OPEC's insurgence could only brighten the nuclear picture yet further. Almost before the first OPEC shockwave had subsided, the US nuclear establishment was pounding on the public's door with the good news. Nuclear power, they declared, would save industrial society from the clutches of OPEC. Week by week US electricity suppliers queued up with order after order for nuclear plants. In the twelve months of 1974 they ordered 26 more reactors, with a total planned output of nearly 31,000 megawatts. To be sure, electronuclear planners encountered irritations in plenty. Many of their projects bogged down in court cases brought by local people opposed to the plants; and some scientists and engineers continued to express concern about safety systems on the plants. But the US nuclear community took a generally sanguine view of its prospects; and so did its opposite numbers in other industrial countries.

Nevertheless, even as the 'nuclear-for-oil' bandwagon was picking up speed, it went over a gaping pothole. On 18 May 1974, underneath the Rajasthan desert, India detonated a nuclear explosive no one knew she had. The Indian explosive was fabricated from plutonium produced in the CIRUS research reactor, given to India by Canada in the mid-1950s and supplied with heavy water by the US. The Canadians were incensed; their nuclear agreement with India expressly excluded using any facilities or material supplied by Canada to manufacture a bomb. The Indians, however, insisted that what made the dent in the desert was not a bomb but a 'peaceful nuclear device'. The semantic distinction did not mollify the Canadians, who forthwith ceased nuclear cooperation with India. The US too made aggrieved noises; but its protest was more rhetorical and less functional.

The Indian nuclear explosion shredded a complacency that had settled about civil nuclear technology, demonstrating with more than ten kilotons of emphasis the possibility that nominally civil nuclear activities could have consequences hard to distinguish from military. Worries about this prospect had been bubbling below the surface of the international nuclear community for some time; the Indian explosion blew them into the open. From that point onward the planned expansion of civil nuclear technology as a 'substitute for oil' fell under a lengthening military shadow, and ran into corresponding diplomatic travail. Governments that had been deploring the activities of nuclear opponents in their countries suddenly found themselves at loggerheads with their diplomatic allies over many of the same issues. Wrangling broke out between the US, Canada, Australia, various European countries, the International Atomic Energy Agency and other international bodies.

The problem was one that the international nuclear community could well have done without. Growing opposition to nuclear developments in many industrial countries was prolonging licensing, delaying construction and imposing costly additional environmental and safety constraints. In some countries court action by objectors brought nuclear programmes to a virtual standstill. At the same time, the original estimates of capital cost of nuclear plants were much lower than the bills that finally arrived;

and the performance of the latest plants regularly fell well short of design specifications. Electricity suppliers began to wilt under the burden of nuclear capital costs; and escalating electricity prices undermined customer demand. As a result, electricity suppliers in most industrial countries found themselves with a growing excess of generating capacity, whose capital cost had to be financed even when the plant was not generating saleable electricity. The dramatic upsurge of nuclear ordering in 1973–4 died away almost as rapidly as it had arisen.

While the nuclear manufacturing industry began to stumble to a standstill, the rest of the so-called 'nuclear fuel cycle' – uranium mining and processing, 'enrichment', fuel fabrication, spent fuel handling and nuclear waste treatment and disposal – fell into ever deeper disarray. Uranium supplies proved to be both economically and politically unpredictable. International disagreements arose between supplier and user governments as to the appropriate measures to prevent a recurrence of the Indian embarrassment. Canada imposed a unilateral ban on shipping uranium to its European clients; the US imposed a similar ban on shipping highly enriched uranium to European research reactors; Australia cut off uranium supplies to France to protest French nuclear weapons tests in the South Pacific. The disputes underlined the close parallel between oil and uranium as politically sensitive commodities; indeed uranium had been so since 1940, predating by some 30 years the 'politicization' of oil.

By the end of the 1970s, and in the aftermath of the accident at Three Mile Island, the prospects for civil nuclear power as a beneficiary of uncertainties associated with oil had become at best speculative. Official forecasts of nuclear capacity in the 1990s and beyond shrank by the month. Reactor manufacturers reported staggering losses, into the hundreds of millions of dollars. They continued to survive on the backlog of orders received before the mid-1970s; but from the early 1980s onward their futures looked distressingly barren; and the brief bloom of nuclear enthusiasm was severely tarnished.

MODEST DOUBTS, EPIC FAILURES

Some fuel-forecasts and plans worked out, to be sure, very well indeed – for instance, as already mentioned in Chapter 3, the British government's heroic decision in the mid-1960s to convert the country's gas supplies from town gas based on coal and oil to natural gas from the North Sea. Some plans crucial to fuel-use appeared initially successful, only to give rise progressively to doubts – for instance the construction of the US interstate highway system, and similar petrol-thirsty road-networks between and even within major cities throughout the industrial world. Until 1973, however, no matter which area of human activity they were considering, government and corporate planners everwhere started from the premise that fuel-use would continue to increase essentially without limit. Moreover, they regarded this as eminently desirable, since higher economic growth went hand in hand with more use of fuels and electricity.

Even in the late 1950s, to be sure, an occasional doubting voice was raised. One prophet crying cogently in the wilderness was E. F. Schumacher, the economic advisor of Britain's National Coal Board. Schumacher demonstrated that the straightforward arithmetic of so-called 'exponential growth' – doubling, redoubling, doubling yet again and so on – that the planners were taking for granted led to wildly implausible numbers on a surprisingly short timescale; but few paid any attention at the time.

In the late 1960s another heretic, M. King Hubbert of the US Geological Survey, made a similar point. King Hubbert investigated all the historical evidence about quantities and rates of oil discoveries, production and use, and the associated geological and economic data. On the basis of his analysis he suggested that the significant role of petroleum in human history would be very brief indeed, probably not much more than two centuries: starting in 1859, peaking early in the twenty-first century and tapering off for perhaps another century. In other words, King Hubbert concluded that the worldwide use of oil, far from continuing to expand indefinitely in double harness with world-wide economic growth, was already nearing its limiting rate of expansion. According to King Hubbert, within a few years new oil discoveries would be unable to keep abreast of the increasing use of oil.

His findings, although greeted with respectful disbelief by most economists, proved harder to ignore than Schumacher's – not least because the intervening years had made Schumacher, too, look a good deal less heretical. However, neither they nor other doubters foresaw just how quickly the position would change. The OPEC oil shock of late 1973 shattered the prevailing consensus of forecasters and planners, based on 'cheap' fuel, almost overnight. What had until recently been called 'fuel and power policy' became 'energy policy'; throughout the rest of the 1970s the policies proliferated. Most of the energy-plans promulgated officially, by governments and corporate groups, shared one attribute in particular. Compared to previous forecasts and plans, they were conceived on an epic scale, and couched in table-thumping rhetoric. They also proved, usually sooner rather than later, to be totally out of touch with the prevailing reality.

One of the first of these grandiose plans was called 'Project Independence'. It was launched with defiant fanfare by US President Richard Nixon at the beginning of 1974, in the immediate aftermath of the OPEC embargo. The declared aim of Project Independence was to ensure that the US could never again be 'held to ransom' by Arab oil-producers. How this was to be achieved remained, however, vague – to put it mildly. President Nixon was already fighting a losing battle of his own, as the sordid saga of the Watergate cover-up unravelled, culminating in his resignation in August 1974. A 4000-page report under the rubric of Project Independence was duly published in November 1974; but it consisted mainly of supply-demand forecasts, with no policy recommendations at all. Indeed it conceded that all the possible options looked unpromising. It declared that the US would inevitably still depend on imported oil by 1980; but that 'incentives' and 'guarantees' could persuade the fuel and electricity suppliers to expand domestic production of oil, coal and especially nuclear power to make the US self-sufficient by 1985. In the event, however, no such incentives and guarantees were forthcoming. A year after Nixon proclaimed it, Project Independence petered out with no effect whatever. US oil imports continued to increase; so did the price of oil.

The European Community, too, was badly rattled by the oil-embargo. All the nine member countries of the Community were more or less dependent on OPEC oil, most of them acutely so. Averaged over the nine, imported oil represented over 60 per cent of the fuel and electricity used – clearly a precarious state of affairs. The bureaucrats of the European Commission took it upon themselves to draft an energy policy for the EC countries, to shake off the OPEC shackles. Knowledgeable insiders subsequently reported that the analysis in Brussels went something like this: 'How much fuel and electricity will the EC require in 1985? How much indigenous oil, coal and natural gas can the EC produce? How much oil will the EC be willing to import? How big is the gap remaining between EC energy requirements, indigenous production and acceptable oil imports? Right – that's how much nuclear energy the EC must produce.'

This back-of-the-envelope cogitation was then expanded into a series of ten policy papers given the Brussels document reference R/3333. Their basic conclusion was that nuclear generating capacity in the EC countries must expand fourteenfold in the decade to 1985. The R/3333 series of papers pursued the ramifications of this extraordinary notion into elaborate detail; but they offered no insight whatever as to how the EC countries might actually achieve the proposed fourteenfold expansion of nuclear capacity in ten years – nor about how much it would cost, or who would pay for it.

The International Atomic Energy Agency (IAEA), as might be expected, also assessed the role that nuclear power might come to play in a world of higher oil prices. By May 1975, the IAEA had concluded that total world nuclear capacity would increase from some 92 gigawatts (GW) – equivalent to 92 large nuclear power stations – in 1975 to 'most likely' capacities of 663GW in 1985, 1350GW in 1990, and 3600GW in the year 2000, with 'maximum' capacities of 287GW, 850GW, 1850GW and 5300GW for the same years. That is, the IAEA anticipated having anything from some 3500 to more than 5000 large nuclear stations in operation worldwide by the year 2000 – starting with the equivalent of 92 in 1975. The analysis was couched in cautious language, warning about the uncertainties involved; as matters turned out, the caution was fully warranted. The IAEA, like the European Commission, could only suggest policies, not implement them; and the practical policymakers proved to view the situation very differently.

FULL SPEED AHEAD IN ALL DIRECTIONS

In November 1974 the OECD published a report on 'Energy Prospects to 1985', subtitled 'An Assessment of Long Term Energy Developments and Related Policies'. To refer to a decade later as 'long term' indicated the difficulties facing forecasters and policymakers; on the existing track record, this so-called 'long term' would be barely long enough to order, design, construct and commission a major energy facility like a refinery or power station. By 1977, when the OECD updated the study as 'World Energy Outlook', its ground-rules had changed appreciably, but its central finding remained broadly the same: 'Faced with the economic and political consequences of increasing (oil) import demand, OECD countries have but one realistic alternative to minimize inherent risks of enlarged imports of oil: to take positive action to expand energy supply – particularly the "conventional" sources of oil, natural gas, coal and nuclear energy; to realize greater energy savings through more effective conservation measures; and to build larger stockpiles to buffer the effect of any deliberate supply reductions.' Although the study nodded in the direction of 'energy conservation', its focus, like those of other studies from official bodies around the world, was on expanding supplies of fuel and electricity, to meet a 'demand' estimated essentially by extrapolating previous trends.

The United Nations, the OECD and the European Community were and are international organizations, whose members are national governments. Although the international organizations can propose policies, only national governments can put them into effect. Governments in turn must reckon with the fuel and electricity suppliers, which include many of the world's largest and most influential industrial corporations – oil companies like Exxon, Shell and British Petroleum, electrical utilities like Electricité de France (EdF), Rheinische-Westfälische Elektrizitätswerke (RWE) and American Electric Power, and engineering companies like Bechtel, ASEA Brown Boveri, Siemens and General Electric, among many others. The oil price shock of 1973 triggered intense activity by the forecasting and planning units of these supplier-companies, each of course endeavouring to stake out the most promising territory for future business.

A synthesis of their early results emerged at the 1977 meeting of the World Energy Conference (WEC) in Istanbul. WEC, founded in 1924 as the World Power Conference, has gradually broadened its scope to encompass not only electricity but also all fuel supplies; and its participants are drawn largely from among practising supply engineers and other professionals. Its analyses therefore tend to be distinctly less rhetorical and more hard-nosed than those prepared by international bureaucrats. But WEC is essentially a suppliers' organization; and it is interested primarily in the future of fuel and electricity supplies – not in how they are used, or what for.

In 1975 WEC established what it called a Conservation Commission, charged to 'evaluate, for the period 1985–2020, future primary-energy supply prospects, and to determine the possible courses of action directed toward overcoming potential shortfalls in energy availability'. The Commission's definition of the 'conservation' in its name was revealing: 'The goal of energy conservation is to achieve acceptable economic growth with a minimum increase in total energy consumption.' However, 'conservation through more efficient utilization of energy must be a key element in future energy policy, but projected future energy demand cannot be satisfied simply by the more efficient use of currently available energy resources. In consequence, a broad energy supply strategy will require all of these features: maximum production of non-renewable resources such as coal, oil, gas and fissile nuclear resources; significant development of non-conventional gas and oil resources; (and) timely development of renewable resources such as hydro, solar, geothermal, biomass and fusion. Furthermore, in the course of the period 1985–2020, extensive substitution of other primary energy resources for oil and gas will be obligatory.'

The WEC 'strategy' thus outlined could be paraphrased essentially as 'full speed ahead in all directions.' Although called the Conservation Commission, the group's attention and its proposed 'strategy' remained fixed firmly on supply of fuel and electricity. Indeed, its summary report concluded with a caution: 'it is well to emphasize that it is always better to err on the conservative side, having a surplus of primary energy supply capability, rather than an energy deficit. Past experience has shown that it is much less costly and socially disturbing to slow down an energy supply programme than it is to accelerate it.' A thoughtful reader might have wondered why, if this were so, WEC was advocating precisely the acceleration that it called 'costly and socially disturbing'. But WEC, as a suppliers' organization, was accustomed to seeing energy policy as a question of supply. WEC started from the presumption that the other side of the equation – energy use – was beyond its control or influence, except by the higher or lower fuel and electricity prices charged. Future energy use was predictable only in aggregates and averages, by extrapolating trends, as an accompaniment to economic growth, adjusted for price-effects but essentially autonomous. It was, in short, an essentially non-negotiable 'demand' from society at large, for which a 'supply' had to be provided. That was the suppliers' job. All society had to do in return was to cooperate with the suppliers: pay the necessary costs, eliminate obstacles like awkward laws and regulations, and offer appropriate incentives – tax breaks, government grants and other inducements. The suppliers would do the rest.

As we shall see, this traditional approach was so deeply ingrained that it continued to dominate official and quasi-official energy thinking for at least another decade. From the mid-1970s onwards, national and international study groups drawn from the top echelons of government, industry and academia grappled with energy policy, reaching conclusions that look frankly embarrassing less than fifteen years later. In 1977, the 'Workshop on Alternative Energy Strategies', a high-powered group from 14 countries sponsored by the Massachusetts Institute of Technology (MIT), began the first chapter of its final report thus: 'After two years of study we conclude that world oil production is likely to level off – perhaps as early as 1985 – and that alternative fuels will have to meet growing energy demand. Large investments and long lead times are required to produce these fuels on a scale large enough to fill the prospective shortage of oil, the fuel that now furnishes most of the world's energy. The task for the world will be to manage a transition from dependence on oil to greater reliance on other fossil fuels, nuclear energy and, later, renewable energy systems.'

The MIT Workshop gave rise to a follow-up, called the 'World Coal Study', with comparably high-powered participants. Its report, entitled 'Coal: Bridge to the Future', published in 1980, started from similar premises and concluded that 'A massive effort to expand facilities for the production, transport and use of coal is urgently required to provide for even modest economic growth in the world between now and the year 2000. Without such increases in coal the outlook is bleak.'

From the perspective of the 1990s, these gloomy prognostications border on incomprehensible. Yet they represent a view espoused in the uppermost circles of official energy thinking not much more than a decade ago. How could they be so comprehensively, wildly wrong?

THE ENERGY POLICY PROJECT

Even before the OPEC shock of 1973, a very different approach to energy issues was taking shape. One of its earliest manifestations emerged in the work of the Energy Policy Project in the US. The Ford Foundation set up the Energy Policy Project in December 1971, appropriating a budget of US$4 million to be spent over three years on a series of studies and reports 'to explore the whole complex of energy issues facing the nation'. The Foundation's decision to launch the

Project was prescient, as OPEC demonstrated less than two years later. In his Foreword to the Project's interim report 'Exploring Energy Choices', published in early 1974, Foundation President McGeorge Bundy noted that 'The problems of energy policy are large and hard, and most public analyses address a limited segment of the problem or argue from the standpoint of a particular interested party'. The Project, by contrast was 'carefully designed at once to avoid control by any special interest and to enlist the advice and counsel of many different kinds of experts' – including not only fuel and electricity engineers, economists, planners and executives, but physicists, lawyers, biologists, environmentalists and writers.

The final report of the Project, 'A Time To Choose', was published later in 1974, supported by a series of some two dozen specialized reports dealing with topics as varied as 'Reclamation of Surface-Mined Land', 'Urban Transport and Energy Saving', and 'Nuclear Theft: Risks and Safeguards'. The Energy Policy Project pioneered an innovative approach very different from traditional fuel-supply fore-casting. Indeed it specifically disavowed any intention to 'forecast' energy 'demand' or 'supply'. Instead the Project team prepared what they called 'scenarios' for future energy-use to the year 2000. Each scenario was designed to be internally consistent, matching supply and use of different fuels and electricity year by year, just like a traditional 'forecast'. Each scenario, however, also acknowledged explicitly a crucial point at best left implicit in traditional forecasts: forecasts influence policies – but policies also influence forecasts.

A traditional forecast is drawn up to direct government or corporate policy, especially concerning investment in supply facilities. Will customers be ready to use and pay for the output from a new mine, refinery, or power plant when it comes into operation some years hence? If so, build it; if not, don't. But what if the government, say, decides to set tighter standards for building insulation or vehicle performance, or to offer tax breaks for replacing inefficient industrial motors with more efficient ones? Such measures will affect future levels of energy use – that is, such policies will affect the forecast 'energy demand'. To a traditional forecaster, working for a supplier hoping to expand its market, measures like these would be beyond control and outside the brief – if not indeed actively unwelcome, since they might reduce future sales of the company's output.

The Energy Policy Project scenarios, however, spelled out all the relevant ground-rules: not only the anticipated prices and supplies of fuels and electricity, but also the framework of government policies and objectives, laws and regulations, import-export constraints and environ-mental criteria to be met. In doing so they identified specific policy measures that could affect future energy use, and analyzed the potential effects. Introducing the approach, the Project's final report noted that

'The three alternate futures, or scenarios, are based upon differing assumptions about growth in energy use. In many ways they are quite dissimilar, but each scenario is consistent with what we know about physical resources and economic effects. The scenarios are not offered as predictions. Instead, they are illustrative, to help test and compare the consequences of different policy choices. In reality, there are infinite energy futures open to the nation, and it is not likely that the real energy future will closely resemble any of our scenarios. Our purpose is to spotlight three possibilities among the many, in order to think more clearly about the implications of different rates of energy growth. What are their effects on the economy, the environment, foreign policy, social equity, life styles? What policies would be likely to bring about each one? What resources are needed to make each of them work?'

As this brisk preamble indicated, the Project extended 'energy policy' far beyond the boundaries of traditional fuel-supply forecasting and planning. Each of the three scenarios was given a label identifying its guiding theme: 'Historical Growth', 'Technical Fix' and 'Zero Energy Growth'. As the label suggested, the 'Historical Growth' scenario extrapolated fuel and electricity use along a curve sloping upward as it had from 1950 to 1970, increasing at about about 3.4 per cent a year, and essentially tripling between 1970 and 2000. The 'Technical Fix' scenario differed little from the 'Historical Growth' scenario in the mix of goods and services provided, but reflected 'a conscious national effort to use energy more efficiently through engineering know-how – that is, by putting to use the practical, economical energy-saving technology that is either available now or soon will be'. This scenario cut the growth rate of fuel and electricity use almost in half – down from 3.4 per cent to only 1.9 per cent, so that the amount used in 2000 would be only two-thirds of that used under 'Historical Growth'. 'Yet the effect on the way people live and work – on material possessions, jobs, comfort, travel convenience – would be, our research tells us, quite moderate'.

The 'Zero Energy Growth' scenario shifted the mix of benefits provided somewhat away from goods and toward services – 'better bus systems, more parks, better health care'. The scenario included all the hardware improvements of 'Technical Fix' plus extra emphasis on efficiency. 'An energy excise tax, by making energy more expensive, would encourage the shift.' The income from this tax would be devoted to the public services, 'to enhance the quality of life, as defined in the scenario'. By the year 2000, the scenario had achieved 'Zero Energy Growth', using only some 60 per cent more fuel and electricity than in 1970, at a rate no longer increasing at all. This proposition left traditional forecasters fuming with disdain and disbelief.

The report, and its backup studies, offered a wealth of data on every aspect of energy use and supply in the US. It concluded with a series of recommendations endorsing the objective of reducing the growth in the use of fuels and electricity at least in line with the Technical Fix scenario, through a wide range of policy measures affecting every aspect of energy. Many of the Project's proposals became recurring motifs in the evolving controversy, as we shall see in later chapters. The report also emphasized that an informed public should participate actively in making and implementing energy policy – and not let the fuel and electricity suppliers alone direct policy by default.

As might be expected, individual members of the Project's Advisory Board took strenuous exception to many of the report's findings and recommendations; fully 62 pages of 'A Time To Choose' were devoted to statements from the advisors. Perhaps predictably, given the recommendation to reduce fuel and electricity use below 'Historical Growth', Project advisors from fuel and electricity suppliers dissented most volubly. On the other hand, advisors with no supply-side affiliations called the report too timid, insisting that it underestimated the environmental impact of fuel and electricity supply and the opportunities for improving efficiency.

By any criterion, the Energy Policy Project was a landmark. It gathered and analyzed information and data never previously integrated into energy planning. It also brought together people whose professional and personal viewpoints collided head-on, making the collisions explicit and revealing the depth of disagreement underlying the confident near-unanimity of energy forecasters hitherto. Where previous energy forecasting had ploughed a narrow, foreordained furrow toward the future, the Energy Policy Project laid out for popular inspection the full range of energy possibilities, inviting active involvement by everyone in society. The questions it raised and the potentials it identified still dominate the energy agenda more than fifteen turbulent years after 'A Time To Choose'.

LOVINS JOINS THE FRAY

The Energy Policy Project in the US was the first attempt to devise an energy policy from the viewpoint not only of fuel and electricity supply but also of energy users, their environment, their government and their country. A yet more striking innovation was already emerging. In Stockholm, in the autumn of 1973 – predating the Yom Kippur war that precipitated the OPEC oil shock – a precocious young American physicist called Amory Lovins presented a paper to a UN scientific Symposium on Population, Resources and Environment. In successive editorial revisions the paper expanded into a book entitled 'World Energy Strategies', which set out a radically different approach to forecasting and planning for energy, not merely for an individual country but for the entire planet.

Trained as a physicist, Lovins was working for a pioneering American environmental campaigner called David Brower, founder of the international environmental organization Friends of the Earth. 'World Energy Strategies' combined hard-headed numeracy, pugnacious combativeness and visionary scope in a heady challenge to energy orthodoxy. The opening paragraph suggests the flavour: 'As medical science, by deferring death, has allowed many more people to live on the earth, so the energy of fossil fuels, by deferring physical scarcity, has kept those people alive. Medical technology has caused a population explosion, energy technology, at least for some, a wealth explosion. How many people can continue to live on the earth for how long and with what wealth depends on the ingenuity and wisdom with which man uses energy. Yet on a planet that is round and therefore finite, energy conversion must eventually encounter some geophysical outer limits; and even sooner, it may be constrained by lack of resources, by biological side-effects, by technical problems, or by social, political and economic pressures.'

That single paragraph – so panoramic compared to the pronouncements of traditional fuel and electricity planners hitherto – flagged the whole vast nexus into which global energy policy has since burgeoned. 'Exploring these constraints and the ways in which they interact with the numbers and goals of people in widely differing societies is arguably the most complex problem in the world today, and one of the most important.' Before setting out on his explorations, however, Lovins appended a puckish warning deflating grandiose pronouncements – including his own: 'Never believe an expert. No expert can tell the whole story, nor avoid emotional entanglements with particular ideas. And never, never believe an expert who is trying to sell you something!'

Unlike the Energy Policy Project, which tried to achieve a measure of consensus among a disparate group of people, Lovins was writing and speaking essentially for himself, and his commentary made his own position emphatically and often abrasively clear. He considered the traditional assumption of indefinitely increasing fuel and electricity use indefensible. Neither the physical resources of fuels nor the financial resources of capital could support unlimited growth in energy conversion, whether fossil or nuclear; nor could the planet sustain the consequent side-effects from extracting, processing and using fuels, and disposing of the resulting wastes.

After a synoptic but incisive survey of the scene, his conclusions were uncompromising: 'The rapid energy growth rates that most industrial countries have long maintained cannot continue for much longer. . . . Most technical fixes that increase energy supply are slow, costly, risky and of temporary benefit; most social or technical fixes that reduce energy demand are fairly quick, free or cheap, safe and of permanent benefit. . . . Governments should suspend their nuclear programmes until enough infallible people can be found to operate them for the next few hundred thousand years. . . . Policy advice should start to come now from those who have tried to prevent our present difficulties, not from those who have caused them but who still dominate energy planning. . . . Planning . . . in an uncertain world must . . . keep as many options open as possible. Present energy policies, however tacit and ill-constructed they may be, are quickly destroying the options that mankind, living and unborn, will need for millenia.'

His aggressive acerbity did not endear Lovins to the traditional energy policy community. But the force of his numerical arguments, and his confident onslaught on the long-standing oracles of the energy business, quickly won him influential allies on both sides of the Atlantic. As Lovins himself was always at pains to point out, he was drawing on the work of many others; but he gave it a cutting edge, sharpening it into a weapon studded with memorable barbs. 'We shall not go so far as to suggest (as some do) that fuel shortages should be relieved by burning energy studies. . . .' 'Living things, intertwined by their individually tiny but collectively vast flows of energy, have been evolving for several billion years within the constraints of the energy income available to them. They work very well and know exactly what to do, even though they have never been to engineering school. . . .' 'Though the price mechanism rightly places a high premium on all the tungsten in China, it also places an exceedingly high premium on live dodos. . . .' Lovins laid about himself exuberantly, belabouring the entrenched forces of the energy tradition.

In October 1976 the US journal 'Foreign Affairs', a high-powered quarterly whose contributors and readers are drawn from the topmost echelons of politics, diplomacy and academia, published a paper by Lovins entitled 'Energy Strategy: The Road Not Taken'. In it Lovins introduced a metaphor that became a touchstone for energy controversy: the 'hard path' versus the 'soft path'. Expanded to book length with the title 'Soft Energy Paths', it became probably the most controversial document to emerge from the new ferment over energy in the 1970s. Some hailed it as a bible for energy utopia; others reviled it as mischievous, misguided and subversive.

The heart of the book was Lovins' premise that in supplying and using energy society could follow either a 'hard' path or a 'soft' path – and that the two paths would be mutually exclusive. According to Lovins, the hard path – essentially extrapolating the US official policy of the mid-1970s – postulated ever-increasing use of fuels and electricity, supplied by ever-larger centralized facilities, converting ever-greater amounts of fuels, particularly coal and uranium, especially into electricity. To Lovins this path posed insuperable problems. It would swallow more capital investment than society could provide, while displacing skilled labour and creating unemployment. It would ravage the environment, both in extracting the fuels and in disposing of their wastes. It would require rigid central control and planning to impose the necessary industrial installations – mines, oilfields, refineries, pipelines, power stations, waste dumps – on localities that might not welcome them.

The feature of Lovins' hard path that drew his fiercest fire was the proposed rapid buildup of nuclear power capacity. Lovins noted that every earlier transition, from coal to oil to natural gas, had entailed moving from a more complex, costly and cumbersome fuel to a simpler, cheaper and more manageable one. The simpler, cheaper fuel entered the energy market with an obvious advantage, carving a share for itself with ease. Nuclear power was a drastic departure from this progression. It was much the most complex of all energy technologies. It had to have a vast array of interlinked facilities – uranium mines and mills, uranium enrichment plants, fuel fabrication plants and power stations – all in place and operating, before it could deliver a joule of electrical energy. These facilities cost staggering sums of capital, which would have to be recovered from electricity users in a fully commercial nuclear power programme. In practice, as Lovins noted, governments – that is, taxpayers – had provided breathtaking sums in subsidy for ostensibly 'commercial' nuclear power, not only in the US but around the world. Taxpayers financed nuclear research and development. Government grants and tax relief fostered uranium mining and milling. Many key nuclear installations had been built primarily for weapons-purposes, out of military budgets, before offering their services at bargain rates for 'civil' applications.

Commercial insurance companies, acutely wary about the scale of possible liability in the event of an accident in a nuclear power station, would provide only limited insurance cover. In the early years of nuclear power, therefore, electricity suppliers refused to build or operate nuclear stations, until governments passed legislation limiting their liability to claims from people affected by nuclear accidents. The status of nuclear insurance, as an indirect but crucial subsidy to this particular energy technology, has been anomalous ever since.

Lovins also sounded an urgent warning about the prevailing assumption that increasing nuclear capacity would soon move from 'conventional' plants burning uranium to so-called 'fast breeder reactors' burning plutonium. In the mid-1970s, in response to a lawsuit led by an independent environmental organization called the Natural Resources Defense Council, the US Atomic Energy Commission had published a massive 'Environmental Impact Statement', weighing more than 5 kilograms, on its plan to have some 400 plutonium-fueled fast breeders in operation in the US by the year 2000. Plutonium is a nuclear explosive, used in almost all modern nuclear weapons. Less than 10 kilograms of plutonium – a lump the size of an orange – is enough to make a bomb as powerful as the one that obliterated Nagasaki. A single large-scale fast breeder power station would require as least 10 tonnes of plutonium for its operation, inside the reactor and in fresh fuel. The Atomic Energy Commission plan would have entailed processing, shipping and storing some 4000 tonnes of plutonium by the year 2000 – enough to manufacture at least 400,000 nuclear bombs – while ensuring that not even 10 kilograms of this 'commercial' plutonium fell into the wrong hands. Even within the nuclear community, doubts were surfacing. Elsewhere, the prospect made knowledgeable observers blench. Lovins was only one of many warning that a so-called 'plutonium economy' was a catastrophe waiting to happen.

Lovins, however, insisted that the 'hard path', with all its unattractive attributes, was in no way inevitable. Indeed it was 'hard' in both senses: not only unyielding but also acutely difficult to follow. On the other hand, society could choose a 'soft path'. Instead of investing in more and larger supply installations, society could redirect capital and effort into improving the efficiency of energy use, and replace the traditional large-scale, centralized supply facilities with many more decentralized, smaller-scale facilities, whose output closely matched users' requirements. Lovins marshalled an impressive accumulation of data from other researchers in many countries to support his arguments; 'Soft Energy Paths' carries a counterpoint of elaborate footnotes on almost every page.

The technical capabilities and economic performance of Lovins' soft path' options were in general amply documented. Perhaps, however, the most immediately telling point, to which Lovins returned repeatedly, was the gaping discrepancy between energy-use in the US and in European countries with similar standards of living. As Lovins emphasized, the US was not notably either more civilized or more comfortable than, say, Federal Germany, Sweden or Denmark; but an average German, Swede or Dane used only about one-half as much fuel and electricity as an average American; and studies in the European countries indicated abundant opportunities for yet higher efficiency.

While still in his mid-twenties, Lovins thus became the most conspicuous embodiment of a radically innovative approach to energy. In subsequent chapters we shall examine the many dimensions of this alternative approach in much more detail. Lovins, as he himself persistently reiterated, was only one of a rapidly growing band of analysts and activists pursuing these exhilarating ideas in many parts of the world. But Lovins was more than an analyst; he was a catalyst, exciting reactions wherever he went, stimulating vigorous exchanges, provoking controversy, demanding that those in power defend their decisions, wrenching energy policy out of the backrooms and into the public arena. He led from the front, bearding chairmen in their boardrooms and challenging editorial writers in their own columns. His very visibility helped to create a focus and a constituency for concepts and ideas hitherto disparate and ineffectual against the concentrated influence of the traditional fuel and electricity supply community.

SWEDEN IN TRANSITION

While Lovins was breaking fresh ground in the US, other analysts were doing likewise elsewhere. In Sweden, a royal commission chaired by the legendary diplomat and internationalist Alva Myrdal published a report in 1972 recommending the establishment of a Secretariat for Futures Studies attached to the Swedish government. One of the Secretariat's first projects was a study of 'Energy and Society', led by Maans Loennroth, with Thomas B. Johansson and Peter Steen. The project's report 'Energy in Transition' was published in 1977, not only in Swedish but also in English. It was circulated to government agencies, members of Parliament, political parties, newspapers, private companies, schools and interested individuals, and eventually published in book form. Unlike 'Soft Energy Paths', 'Energy In Transition', as the product of a government group, had an official imprimatur as a discussion document. It was accordingly more measured and less pugnacious than the Lovins book; in many respects, however, it was fully as radical – especially given its official origins. It was also, even in English, easier to read than 'Soft Energy Paths', whose arguments tended to emerge not only in fusillades of numbers but also in uncompromising Mandarin prose; indeed Lovins' wife Hunter was once heard to declare that one of her projects would be 'translating "Soft Energy Paths" into English'.

As the title of 'Energy In Transition' indicated, Loennroth's group viewed the energy issue somewhat differently from Lovins. Lovins insisted that the 'hard path' and 'soft path' of energy development were mutually incompatible, and that choosing one would foreclose choosing the other. Loennroth's group likewise identified two possible directions akin to those of Lovins, and agreed that one might eventually exclude the other. But they regarded the immediate future as a period of 'transition', in which the objective of energy policy would be to maintain what they called 'freedom of action' – freedom to pursue either of the two directions. Beset by every kind of uncertainty – about fuel supplies and prices, technical questions, and short-term and long-term environmental and social issues – prudent policy-makers would be unwise to reject either direction prematurely.

In Sweden, with no domestic fossil fuels, they saw one direction as based ultimately on imported coal and/or nuclear power, including fast breeder reactors; and the other as based on 'renewable' sources like solar and wind energy and so-called 'biomass'; see Chapter 8. Loennroth's group pointed out that the first direction was essentially a continuation of current Swedish policy, dating back to the 1956 report of the Fuels Commission, which had recommended starting a nuclear power programme in Sweden. But Loennroth's group argued that true 'freedom of action' meant that Sweden's political policy-makers in Cabinet and Parliament should also be able if they wished to pursue the second direction, toward reliance on renewable sources.

The momentum of existing policy, however, was steadily foreclosing this second direction. Until the mid-1970s, nuclear and to some extent coal technologies had received almost all available energy research and development funding; renewables had received very little. Organizations involved in nuclear power were already well-established and influential; renewables had no comparable organizational backing. Energy users were installing equipment like electrical heating for households, that would make a changeover to renewables more difficult; and so on. If the Swedish government wanted to retain genuine freedom to change the direction of Sweden's future energy development, it would have to take positive measures to strengthen the second option – the alternative, renewable option. Otherwise nuclear power and imported coal – no matter how serious their problems – would be Sweden's only long-term energy options.

Although presented in low-key tones and supported by scrupulously qualified arguments, this conclusion in an official report was political dynamite in Sweden in 1977. Only a few months earlier Swedish voters had thrown the country's Social Democrats out of office after some 40 years in power, in a general election whose hottest controversy had centred on Sweden's nuclear power programme. Social Democrat leader Olaf Palme remained a firm supporter of Sweden's nuclear plans; but his Social Democrats were deeply divided on the issue. The newly-elected Prime Minister, Thorbjorn Faelldin of the Centre Party, was an outspoken opponent of nuclear power, and was in no doubt that this policy had made him Prime Minister. But Sweden's industrial leaders were equally outspoken advocates of nuclear power, and in no mood to back down. 'Energy In Transition' was a moderate, understated document; dropped into this seething cauldron of controversy, however, it could not fail to make an impact.

In January 1979 the Secretariat for Futures Studies published a further, much more detailed report by the same authors, entitled 'Sweden Beyond Oil: Nuclear Commitments and Solar Options'. It described two alternative pathways into the twenty-first century, one based on nuclear energy and the other on solar energy, and discussed the technical, economic, social and environmental implications of these two alternatives. When the report was first published, the Swedish government, by now a coalition, was about to authorize a further stage of the Swedish nuclear programme. Two months later, however, on 28 March 1979, the Three Mile Island nuclear power plant in Pennsylvania suffered an accident that verged on catastrophe. By the time the Swedish report appeared in English translation a year later, under the title 'Solar versus Nuclear: Choosing Energy Futures', the Swedish scene had changed dramatically. In the aftermath of Three Mile Island, the Swedish government abruptly changed its stance, and agreed to hold a referendum on the future of nuclear power in Sweden.

By the time of the referendum, on 23 March 1980, the streets of Stockholm thronged with supporters wearing the distinctive lapel badges of the three referendum choices. One choice was to shut down all Sweden's reactors within ten years, and build no more; one was to complete the existing reactors, operate them, shut them all down by the year 2010 and build no more; and one was to do likewise, but take all the reactors into state ownership. If you consider those three options again carefully, you will notice a curious and startling omission. Although Sweden's political parties wrangled at length over the precise wording of the three lines on the referendum ballot, they eventually agreed on three options all of which terminated Sweden's nuclear programme by 2010. Even before the referendum was held, Sweden's politicians had agreed to shut down all existing reactors by 2010 and build no more.

In the event, voters rejected the short-term shutdown option; but from 1980 onward Sweden had set itself on a course to turn away from the nuclear option and establish an alternative direction – and to have the alternative fully functional no later than 2010. 'Solar versus Nuclear' had come down 'solar' – in the northern climes of Sweden, of all places. 'Energy In Transition' would be a fascinating and illuminating spectacle, for Swedes and non-Swedes alike, as we shall see.

'ENERGY ANALYSIS'

Across the North Sea to the west of Scandinavia, Britain represented an almost complete contrast to Sweden, at least in energy terms. Unlike Sweden, Britain had vast reserves of coal; it also had offshore North Sea oil and gas. The title of a 1975 book by one of Britain's pioneering energy analysts, Peter Chapman, described Britain neatly if quirkily: 'Fuel's Paradise'. Chapman was director of the Energy Research Group at Britain's unique Open University (OU), whose far-flung student body was tutored largely by correspondence, radio and television. The academic staff of the OU Faculty of Technology did not have access to the kind of expensive centralized laboratory facilities possessed by traditional universities. Instead, the Energy Research Group (ERG) initiated a novel discipline that came to be called 'energy analysis'. Its object was to assemble and dissect data on the energy conversions associated with every aspect of human endeavour: running households and farms, extracting and processing primary minerals and other materials, constructing and operating buildings, manufacturing goods of every kind, transporting commodities and people – in effect a comprehensive and detailed physical account of human energy use.

ERG's efforts to tease out and correlate the available data soon gave rise to a technique called 'energy accounting'. It was analogous to financial accounting; but instead of tracking cash flows and financial investments, it tracked energy flows and investments, in particular those controlled by people. Consider, for instance, the light by which you are reading these lines. If you are reading by daylight, the light energy is arriving without human intervention. If, however, you are reading at night, you are relying on a system built and operated by humans. To build and operate the system they use energy. They use, for instance, so much energy to extract a tonne of iron ore; so much energy to produce a tonne of steel from the ore; so much energy to fabricate the steel into girders, reinforcing rods, wires and such; and likewise for concrete, plastics, fabrics and all the other materials that go into a power station. These are 'energy investments': the energy is used to create a long-lived structure.

Having built the stations, transmission lines and such, they then require a continuous input of the appropriate fuel – for instance coal. Miners use so much energy to extract a tonne of coal from the ground and to deliver it to the power station; electrical engineers use energy to process the coal and feed it into the boilers, to burn it, raise steam and generate electricity. Furthermore, the mining machinery is made from steel and other materials – which are yet other energy investments; and so it goes. The system that lights the page you are reading involves a network of interlinked energy flows and energy investments, of Byzantine complexity. The work of Chapman and his colleagues helped to disentangle this network, enabling people to understand much more clearly not only where the energy was coming from but where it was going. Chapman's book 'Fuel's Paradise' was an idiosyncratic but thought-provoking popular guide to the field. It was also a seminal example of how to devise an 'energy policy' starting not with fuel and electricity supplies, but with how people use energy, and what for.

A controversy over using 'energy accounting' for nuclear power spurred widespread interest in energy analysis as a useful discipline, casting fresh light on how we actually use energy – not in terms of broad averages and aggregates but stage by stage in specific processes. A study entitled 'Energy and Food Production' by Gerald Leach, published in 1976, demonstrated vividly the power of this discipline as a tool for policy-makers. Leach, one-time science correspondent for the London 'Observer', combined an incisive and skeptical mind with a lucid and dispassionate prose style; he was to become an energy analyst of international stature.

Leach warned that 'energy analysis is a complicated game, riddled with numbers, conventions and references'; for instance, the 'energy input' to an activity using a litre of diesel fuel 'is not only the gross heat content of the fuel, but an estimate of all the fuels consumed in exploring for oil, extracting it from the ground, shipping it, refining and delivering it to a final consumer, including that used to provide all the materials for and to build all the machinery and plant employed in the entire production delivery chain' – a catechism that might stop an average diesel user from ever starting up his tractor. But the main body of Leach's study offered fully 85 'energy budgets' for food-producing activities. They included whole UK farms of different kinds; individual crops; bread, potatoes and beet sugar; milk; poultry and eggs; vegetables; and other UK agricultural products; and – for comparison – other food-gathering and agriculture, from Kalahari Bushmen to Hong Kong rice-growers to Peruvian anchovy fishermen.

Leach left readers in no doubt about the policy implications of his findings. 'No technical fix will alter the inequity that allows, for example, 55 million Britons to consume enough primary foodstuff to feed half the Indian subcontinent, or to use almost a tonne of oil per head in doing so. . . . The food supply systems of the rich, represented here by the UK, have become heavily dependent on large injections of fuel subsidy. While this energy prop has improved the productivity of the land and of labour, the overall improvement when one looks at the entire food system has not been all that dramatic . . . the apparent "success" of these micro-improvements and the dominance of Western advice and technology . . . has persuaded much of the world that the Western way is also the best development path to follow.

'Fortunately, a new vision that other development paths are both more promising and more practical has now begun to seize the minds of scientists, planners and politicians throughout the developing world. Carbon copies of Western methods are increasingly seen as irrelevant due to their high cost, high resource use and labour-saving characteristics, and their persistent tendency to widen the development gap between the cities and the countryside, the haves and the have nots. Instead, the new emphasis is on decentralized rural development using appropriate small-scale technologies in the context of self-help and self-reliance.' As we shall see in later chapters, the argument is yet stronger a decade and a half later – but may not be strong enough.

Within a few years Leach was to be involved first-hand in the problems of rural Third World energy use that he had signposted in 'Energy and Food Production'. In the late 1970s, however, he and his colleagues at the International Institute for Environment and Development in London prepared and published a report that reverberated through the British energy scene and sent ripples all over the world. Published in January 1979, it was entitled 'A Low Energy Strategy for the UK'; and its opening page set the stage for a remarkable thesis. 'There is widespread pessimism about the energy future. Forecasts show energy needs rising implacably, with widening energy gaps appearing around the turn of the century as oil and gas production begin to decline. They show nuclear power having to be expanded at unprecedented rates to close the gaps and capital being drained away from other uses in the effort to keep energy supplies in step with rising demand.

'This book presents a different view of the future. It does so for the United Kingdom but its approach and findings should hold broadly for other industrial countries. It demonstrates, systematically, and in detail, how the United Kingdom could have fifty years of prosperous material growth and yet use less primary energy than it does today.' Think about that startling claim. It overturned in a sentence all the ruling precepts of traditional energy planning, and indeed carried the fray onto terrain hitherto usually conceded to the traditionalists. Could we really use 'less primary energy' – that is, fuels and electricity – and experience not a 'change in lifestyles' but 'fifty years of prosperous material growth'?

'Such a conclusion upsets the conventional view that a low energy future must be bleak and repressive. It challenges the technical optimists of the nuclear and other energy supply industries who claim that they can bridge whatever energy gaps arise, and suggests that they are the true pessimists. They are anticipating and endeavouring to solve problems which need not arise. There are simpler, safer and cheaper ways of dealing with our energy future than they propose. Our conclusion also challenges the social optimists who believe that we can avoid an energy crisis by fundamental changes in attitudes and practices; whether nobly, by turning from acquisitiveness to altruism; austerely, by wearing pullovers and bicycling; or technically by covering the place with windmills, solar collectors and fuel farms. Although such changes may be desirable they are not necessary.

'We show that Britain – and by implication other countries – can move into a prosperous low-energy future with no more than moderate change. All that is necessary is to apply with a commitment little more vigorous than is being shown today by government, industry and other agencies some of the technical advances in energy use which have been made, and are still being made, in response to the oil price increases of 1973/74.' And show they did, with a thoroughness and attention to detail hitherto unprecedented.

They began by demolishing the traditional assumption linking energy use with 'gross domestic product' (GDP) – the sum total of all financial transitions taking place within a country's borders. Traditional energy analysis started from the premise that economic growth must be paralleled by growing use of fuels and electricity; yet Leach and his colleagues demonstrated with ease that in many areas of energy-use the energy-GDP correlation was inherently unsound. In the UK the non-manufacturing sector 'generates over 50 per cent of the total GDP but it uses only 12 per cent of the total energy'. GDP here depended not on fuels and electricity but on 'the skills and activities of human beings'. Fuel and electricity use was 'mostly for heating and lighting buildings and could easily fall while GDP rose substantially'. For instance, an elaborate analysis showed that in UK housing, fuel and electricity use had grown in direct proportion to income for the past 25 years. 'But this tells us nothing about the income level at which houses became so hot that they are uninhabitable': indeed it implied that the more you earned, the hotter your house. Leach subsequently called this dubious assumption the 'roast-the-rich' theory.

Instead, 'Our forecasts of possible energy futures do not attempt to extrapolate the trends of the past through the discontinuity of 1973/74. They are almost entirely forward-looking; and detailed in building up the total use of energy like a wall, brick by brick. They are designed to detect saturation effects' – like the uninhabitable house of the 'roast-the-rich' theory – and other related limitations, like those on substituting one fuel for another for specific end uses – remember the gas-operated radio of the 1930s. 'Essentially the approach is to start wherever possible with the ultimate purpose for which energy is used – the useful energy demand – and work upwards from there to primary energy supplies, fuel by fuel and subsector by subsector', using the 'many excellent studies' carried out after 1973 'showing in detail how different fuels are used in different sectors to provide useful energy. Using these studies we were able to break down energy demand in 1976 into nearly 400 separate categories determined by end-users, fuels and appliances.'

They then identified an array of activities, and associated each with its use of fuels and electricity. For housing, for instance, they considered 'average interior temperatures, dwelling size and volume, quantity of hot water used per person, the amount of cooking, and the ownership of seven categories of electrical equipment'. They then tracked these multifarious activities and their associated energy use to the year 2025, assuming 'business as usual, but more of it. . . . Houses become warmer, so that everyone enjoys the amenities of the better-off today. Most families come to own freezers, dishwashers, clothes driers and colour TV, and other heavy users of electricity.' Car ownership and traffic, and air traffic, both grow rapidly. The report notes that 'this may be dismaying to those opposed to such growth'; but 'our choice of a fairly high growth future deliberately avoids such issues ... to keep the arguments focused upon energy policies. . . . We wanted to depart as little as possible from the basis on which official energy forecasts are made. . . . We have not supposed a future of material austerity. Nor have we assumed implausibly rigorous conservation.' Nevertheless, Leach and his colleagues then dropped their bombshell. Using all these cautious assumptions, and building up the total picture of energy use sector by sector, they found that in their 'High' scenario that total primary energy – fuels and electricity – use in the UK in 2025 would be not higher but 8 per cent lower than that in 1975. In their 'low' scenario it would be 22 per cent lower.

In subsequent chapters we shall discuss in more detail the analytic approach leading them to this startling conclusion; it was to become the key to the energy alternative now gaining momentum worldwide. Many other analysts in many other countries were by this time developing and applying similar techniques, reaching similar conclusions and publishing their findings. In 1979 the Energy Project at the Harvard Business School, not usually considered a hotbed of radicalism, produced a report entitled 'Energy Future', edited by Robert Stobaugh and Daniel Yergin, pointing in the same direction for the US as the Leach report for the UK. Other similar reports followed, in the US and many other industrial countries; see Further Reading.

THE TRADITION COUNTERATTACKS

Faced by this mounting challenge not only to their policies but also to their authority, traditional energy planners launched an all-out counter-attack. In Britain, a few months after the Leach report appeared, the government's Department of Energy published Energy Paper 39, a two-volume report on 'Energy Technologies for the United Kingdom'. It was prepared by a group known as ACORD, usually spelled out as the Advisory Council on Research and Development; but its full title was actually the Advisory Council on Research and Development for Fuel and Power – and this accurately characterized both its participants and its orientation. Energy Paper 39 was a follow-up to Energy Paper 11, on 'Energy Research and Development for the United Kingdom'; both employed similar analytic techniques and both reached similar conclusions. The approach adopted the notion of 'scenarios' governed by different leading assumptions. Unlike the Leach study, however, the Energy Paper scenarios were generated in the traditional way, essentially by extrapolating aggregates and averages, adjusting for various correlations between fuel and electricity prices, economic growth and energy use. The papers declared that the three key elements of a sound strategy were conservation, coal and nuclear power – but offered only minimal detail about distributing effort among these elements. This 'coconuke' package, as it came to be called, amounted in practice to endorsing existing priorities and policies, notably the egregiously topheavy official support for nuclear research and development, which routinely received hundreds of times as much government money as any other energy technology.

While questioning the economic and social assumptions of 'low energy futures' like those proposed by Leach and others, the Energy Paper authors did not, apparently, feel any parallel need to ask whether their own 'coconuke' scenarios might raise similar questions – perhaps even more acutely. Instead, in media discussions in the months after the Leach study was published, senior figures from the fuel and electricity supply industries uttered stern monitions about the need for 'coercion', and even the establishment of 'energy police' to enforce their version of a Leach-type energy strategy. Coming from industries that invoked compulsory purchase of land, and unilaterally disconnected tens of thousands of customers every year for tardy payment, a warning about 'coercion' was double-edged, to put it mildly.

This controversy in the UK was but one example of a dichotomy rapidly widening and deepening in industrial society almost everywhere, between the energy tradition and the energy alternative. Traditional analysts, focusing on plans to augment fuel and electricity supplies, had a constituency and a power-base built up over many decades, especially in the major supply industries. Innovative analysts, by contrast, had entered the scene only in the late 1960s. Moreover, since they were focusing more and more not on fuels and electricity at all, but on the many different end-uses of energy, their constituency was similarly scattered and various; nor could they themselves actually implement the policies they proposed. Unlike traditional analysts based in governments and corporations, innovative analysts based in universities and environmental organizations had no automatic access to the political levers of power. But the innovators were building up an international network of their own. They exchanged bulletins, reports and commentaries; they met at conferences and seminars; they swapped basic technical and economic information, and debated its validity and significance; and they assembled and disseminated a swiftly-growing body of experience on the alternative approach to energy use and supply.

Traditional energy analysis meanwhile moved into yet higher gear. The International Institute for Applied Systems Analysis – IIASA for short – was a remarkable institution set up in 1972 in Laxenburg, Austria, under the joint sponsorship of the Academies of Science of the US, the Soviet Union and other countries, with the blessings and financial support of their governments. IIASA brought together top experts from East and West to pool their efforts and confront common problems. IIASA's first major study, on energy, expanded into its Energy Systems Programme, involving more than 140 scientists from some 20 countries, under the direction of Professor Wolf Haefele, a senior nuclear-research administrator from Federal Germany. The IIASA programme published its final report in 1981, under the title 'Energy for a Finite World'. It was by any criterion a massive exercise, involving elaborate computer modelling of many different aspects of the global energy scene. It concluded that by the year 2020 the total amount of fuels and electricity used worldwide would nearly double, and might nearly triple. By the year 2030 it might nearly quadruple.

The IIASA study could be considered the apotheosis of the traditional approach to energy. Nevertheless, amid the tidal waves of paper emerging from computers running overtime, a skeptical observer was compelled to keep in mind an acronym coined in the early days of computing. In the 1950s, when a mainframe glowing with vacuum tubes filled a building, and drew enough electricity to power an industrial estate, disgruntled programmers knew that no matter how powerful the computer, it could process only the data entered. A really egregious example they would call 'GIGO': 'garbage in, garbage out'. The IIASA study was not exactly GIGO: but its conclusions followed inevitably from its premises, as ex-IIASA staffers Brian Wynne and Robert Keepin demonstrated in a devastating demolition job published in the British journal 'Nature' in 1984. The assumptions built into the initial structure of the study took a convoluted route through the electronic crank-turning of the mathematical equations; they emerged transfigured and unrecognizable; but they were still the initial assumptions, and they were still based on the traditional relationships between economic growth and energy use. However heavily qualified, they were still essentially extrapolations of aggregates and averages; and they still focused on supplies of fuels and electricity – not on what people actually did with them.

Two years later, in 1983, the World Energy Conference (WEC) convened yet again; and yet again it pondered a set of projections postulating the usual twofold increase in global use of fuels and electricity by 2020. Against a background of collapsing prospects for nuclear power, increasing concern about air pollution and acid rain from fossil fuels, and precarious instability of prices as a basis for long-term investment, the WEC prognosis looked more implausible than ever.

ENERGY FUTURE: DESTINY OR CHOICE?

For more than a decade the confident official forecasts of fuel and electricity prices, supply and use had proved in the event to be ever wider of the mark. By the mid-1980s official forecasters and their political mentors were prefacing each prediction with a disclaimer: 'Forecasts are always wrong, including this one.' The traditional approach to energy planning, expanding supplies of fuels and electricity – 'supplying demand because people demand supply' – was in serious trouble.

By this time, however, another major study was under way, which was to take a very different view of the energy issue, and reach very different conclusions. It was directed by leading energy experts from four continents: Jose Goldemberg from Brazil, Thomas Johansson from Sweden, Amulya Reddy from India and Robert Williams from the US; and it drew together the work of innovative energy analysts from many different backgrounds and affiliations. The study began under the unendearing rubric of the 'End-use Oriented Global Energy Project'.

But its final report, published in 1988, was to have a much more appealing title: 'Energy for a Sustainable World'. The study reached a remarkable conclusion. According to Goldemberg and his colleagues, within 40 years everyone on earth could have the material well-being now enjoyed in industrial countries, while global use of fuels and electricity increased by only ten per cent.

The preface to the report concluded with the following words: 'Our new approach has led us to replace the perspective proposed in more conventional energy analyses by a vision of the global future that is dramatically different. A world adopting the energy strategy we describe would be more equitable, economically viable and environmentally sound. It would also be more conducive to achieving self-reliance and peace. And it would offer hope for the long term future.

'This book suggests that, contrary to widely held beliefs, the future for energy is very much more a matter of choice than of destiny. Energy futures compatible with a sustainable world are within grasp. The choices we urge require imaginative political leadership, but they represent far less difficult and hazardous options for this leadership than those demanded by the conventional projections of the world's energy future.

'Our work is only a preliminary effort to work out the implications of the new approach. Many other efforts are needed. The joy in this endeavour comes from the feeling of being harbingers of hope rather than prophets of doom.'

Brave words: are they warranted, or naively over-optimistic? In the following chapters we shall describe this alternative approach, and explore its implications – beginning with its central paradox. To provide 'energy for a sustainable world', you start from the premise that no one really wants energy anyway.

6
WHO WANTS ENERGY?

All the fuss about energy overlooks one crucial fact. No one actually wants 'energy'. We want comfort, no matter what the climate or the weather. We want light when the sun is not shining. We want cooked food – more nourishing, more varied and easier to digest. We want hot water, to clean and refresh our bodies, our clothes and our utensils. We want to extract and process materials like metals and fibres. We want tools and other hardware, to make work easier. We want to move heavy loads. We want to travel from place to place; and so on. In modern society we satisfy these desires by using energy. What we want is not the energy itself, but the SERVICES that using energy can provide.

USING ENERGY

In Chapter 3 we surveyed human uses of energy as they evolved from about 1800 onwards. By the 1990s the proportions of the picture have changed almost beyond recognition; but the basic physical objectives we seek to achieve by converting natural ambient energy and fuel energy are not as different as you might expect. Consider the 1990s household of John Londoner's great-great-grandson, Peter. Peter Londoner and his family live in a comfortable house in the middle of a terrace in the suburbs. What energy hardware do they use? Their house itself, among its other attributes, is a key form of energy hardware, although they might not recognize it as such. Within the house are electric light bulbs and fluorescent lamps; gas central heating, a gas cooker and a gas heater for hot water; and a portable electric fan heater. In the kitchen are an electric refrigerator and deep freeze; nearby is an electric washer-drier for clothes. About the house are a number of portable radios, operated either from electric batteries or by plugging into the mains electricity supply. In the living room are a stereo and a colour television. In Peter Londoner's study is a computer word-processor, and on his workbench an electric drill. The household also boasts a vacuum cleaner indoors and a power mower outdoors. In the garage are two cars.

Quite an impressive array of assorted energy hardware, you might think – a substantial change from the limited facilities available to his great-great-grandfather. However, the specific physical functions of this array of hardware differ less than you might realize from those accomplished in John Londoner's household almost two centuries ago. John Londoner's family obtained warmth, lighting, cooking and hot water by burning fuel directly. However, they obtained cooling – if it was available at all – not by converting fuel-energy but by filling an insulated box with ice gathered in the winter. Nor did John Londoner have the services of motors, either fuel-burning or electric; nor of course did he have the services of electronics. These three physical functions – cooling, motive power and the subtle manipulation of electrons – are the major additions to the energy services Peter Londoner can call on.

Peter Londoner and his family regard travel very much as a matter of course. They travel for work and for leisure: by car, by public transport buses and underground, by train between cities, and by plane between countries and continents; some day they look forward to a leisurely journey by ship purely for pleasure. Converting fuel energy in motors and engines is an inherent part of the process by which they get from A to B.

Peter works in an office block built in the 1960s. It is sheathed entirely in glass; but its windows cannot be opened. It is heated and ventilated from a large fuel-burning installation in its basement. The lights in the office block cannot be switched on and off individually. The building has high-speed lifts, and a large room filled with computers. The building's lavatories are supplied with hot water, as is the cafeteria on the first floor, which is also equipped with a row of ovens. Peter, of course, pays no attention to the energy conversion processes required to run the office block, except that he finds the relentless hum of the ventilation system irritating, and the blank brightness of the overhead lighting hard on his eyes.

Peter and his family get their everyday food, drink, metals, minerals, fibres and other materials from all over the world, without thinking. They take for granted the substantial energy conversions needed to transport, process and package the goods they acquire. Indeed most of the energy costs that Peter and his family incur are invisible to them. Despite the size of their fuel and electricity bills, they simply do not recognize how much their other household bills depend on energy costs arising somewhere else.

Until the mid-1970s, almost all available data on the role of energy in society was gathered by or on behalf of fuel and electricity suppliers, in the categories that interested them. They could say how much coal was used, how much oil, how much gas, how much electricity, what types and where it came from, who used it, how much it cost and so on. But they gathered surprisingly little detailed information about what all these energy carriers were used in and for: the energy characteristics and performance of buildings, vehicles, boilers, furnaces, lights, motors, electronics and all the other energy hardware upon which modern industrial society depends.

ANALYSING ENERGY SERVICES

Consider, for instance, your house. A house is, among other things, an energy system. Indeed you could start with a box in the middle of an open field. Even this primitive house has a more stable temperature inside than outside, because the box impedes the movement of the air. If you sit inside the box, its walls will slow down the dissipation of your body heat, and keep you slightly warmer than you would be outside the box. Alternatively, if the box were on the Sahara sand, it would be slightly cooler inside, because the walls would keep out the energy of the sun's rays. A more sophisticated house achieves much the same objectives. It stabilizes the indoor temperature, at a level closer to that which the occupants find comfortable. As a sophisticated box, a house lets you intervene in ambient energy conversion. Daytime, night-time, summer, winter and the weather will be there whether you are indoors or out; but the thermal environment indoors may be more congenial on average, especially when conditions outside deviate too far from what you desire.

The more the temperature outdoors differs from that indoors, the greater the challenge to the house or other building. Suppose, for instance, you want to keep the indoor temperature at 18 degrees Celsius. If the average temperature outdoors is 17 degrees Celsius for 24 hours, your building has to compensate one degree for one day: one 'degree-day'. If the outdoor temperature stays constant at 17 degrees Celsius and if your building is unoccupied by people or other energy systems, heat will flow from indoors out: eventually the temperature indoors will also be 17 degrees Celsius. However, a healthy adult, even when sitting down, converts energy at over 100 watts, and at a temperature of about 35 degrees Celsius. Non-biological energy systems – lights, appliances, electronics and such – may likewise contribute higher-temperature heat indoors. The building structure acts as a 'heat reservoir'. If the structure is colder than its surroundings, it soaks up heat; if the structure is warmer than the surroundings it gives off heat.

The more substantial the building, the more it tends to stabilize the temperature inside it. If the temperature outside is 10 degrees Celsius above zero, a tent and a modicum of clothing can keep you comfortably warm. But if the temperature outside is 10 degrees below zero, converting food energy in your body will not be able to keep up with the heat loss through the tent; you may well freeze to death. To forestall this eventuality, you can don arctic clothing and get into a sleeping bag; or you can light a fire. The same alternatives apply, by analogy, to a structure more permanent than a tent. Meteorological records indicate how many degree-days a building may have to compensate at a particular location over a 12-month period. Thoughtful architects and engineers set about investigating more closely the balance between the building structure itself, as a converter of ambient energy, and the heating system requiring some energy carrier from farther away. As we shall discuss in the next chapter, what they learned was both revealing and encouraging.

In most populated regions of the earth, people can stay comfortable by adjusting their surrounding temperature up or down by at most a few tens of degrees Celsius. This physical function is much the most important category of energy use we control; it enables us to survive as living organisms all year round, over most of the earth. As a physical process it is also comparatively uncomplicated and undemanding. To adjust temperatures over such a modest range, you can choose from a vast range of measures, from thicker sweaters to loft insulation to nuclear-powered electric fires. Heating water is physically somewhat more difficult, because you may want to achieve a temperature as high as the boiling point – 100 degrees Celsius; and cooking food may require temperatures of up to 220 degrees or even higher. To establish and maintain a temperature this high, you have to transfer heat energy into the water or the food faster than energy is escaping. If you want to boil a kettle quickly, of course, you put a lid on it, so that the heat energy entering the water from the flame below or from the heating element inside does not rapidly leave again at the upper surface of the water. An oven door must seal tightly and not be opened unnecessarily; oven walls must be thickly insulated to keep the heat energy inside. The higher the temperature inside, the more rapidly the heat will escape to the cool exterior, and the more rapidly you have to replace the lost energy by supplying more. Here again, as with the house, the detailed design and operation of the hardware makes a major difference to how much energy you have to deliver from fuel or electricity to heat your water or cook your food.

You can get comfort, hot water and cooked food by adjusting local temperatures over a few hundred degrees Celsius, by controlling heat energy with appropriate hardware. You can also provide light energy by using heat energy. Just heat a material to a temperature so high that it gives off visible light energy, like a flame, or a glowing coal, or the tungsten filament in an incandescent light bulb. Inevitably, however, if you use heat to generate light, much of the energy you supply escapes invisibly to the surroundings: hold your hand above an incandescent bulb and feel the invisible wasted energy leaving. You know better than to seize a lighted incandescent bulb; but you can hold a lighted 'neon sign' or 'fluorescent' lamp with ease, because these operate essentially at room temperature. An electric current flows through a low-pressure gas inside the bulb, turning the gas molecules into charged particles. In a neon sign the gas then glows, with a characteristic colour – red for neon gas, white for argon gas, and so on. The glass of a fluorescent lamp is coated on the inside with a material called a phosphor; the charged gas-particles make this phosphor glow. Since neon signs and fluorescent lamps operate at room temperature, they lose very little energy as invisible heat. You might not want to light a room with a neon sign; but a fluorescent lamp can light a room using much less electricity than an incandescent bulb of the same brightness.

You can raise the local temperature easily, simply by burning fuel; but lowering the local temperature by using energy is trickier. Refrigerators, deep freezes and air conditioners all generally use a 'heat pump': a device that will collect heat energy at a lower temperature and eject it at a higher temperature. A good refrigerator or deep freeze involves two components: a well-insulated box, to impede the flow of heat from the warm outside to the cold inside; and a heat pump, to extract energy from the inside. Reach behind your refrigerator and feel the heat it is pumping out of its interior. An air conditioner pumps heat out of an entire room, or even a whole building; but the principle is the same. The heat pump needs an appropriate 'coolant', to carry the heat away, and a pump to circulate it. Until the late 1980s, the preferred coolants were inert chemical fluids called chlorofluorocarbons, or CFCs. CFCs had all the desirable cooling properties, were neither toxic nor explosive and seemed as benign as chemicals could be. Now we know, alas, that CFCs are damaging the atmospheric ozone layer that protects the earth's living organisms from the sun's dangerous hard ultraviolet radiation. The search for alternative coolants is one of the least-expected problems now facing energy engineers. In any event, an appropriate coolant can be circulated by an electric motor, or by a gas-flame that sets up a current, or even by the energy of sunlight. 'Solar cooling' is an unexpected and appealing example of 'energy judo': the hotter the sun, the more effective the refrigerator or air conditioner, turning an apparent disadvantage into an advantage.

Another important energy service is the physical function of moving massive loads or lifting heavy objects – that is, of exerting forces and doing mechanical work. Steam engines, internal combustion engines and jet engines all variously convert heat energy from burning fuel to produce mechanical force and movement; electric motors make energy conversion processes yet more flexible. Heat engines inevitably lose a significant proportion of the fuel energy – often well over half – as wasted heat, that escapes with the exhaust gases and from hot engine surfaces into the surroundings. Electric motors, by contrast, can convert well over 90 per cent of the incoming electrical energy into mechanical movement. But if the electricity has to be generated by a heat engine like a steam turbine or a diesel engine, and transmitted some distance over wires, the net useful mechanical energy delivered may still be less than 30 per cent of the original fuel energy. Moreover, engines of whatever type usually deliver mechanical energy by turning a shaft or driving a piston back and forth; friction in bearings, linkages and gear-trains can waste a lot more of the original energy, by turning it into low-temperature, low-grade heat lost to the surroundings.

If an energy-conversion process delivers, say, 30 per cent of the original input energy as final useful energy, engineers say that the process has an 'efficiency' of 30 per cent; and so on. In general, the higher the efficiency the better; doubling the efficiency means that the system can provide the same service while using only half as much fuel or electricity. But energy hardware able to operate with higher efficiency may be more expensive, or raise other difficulties. As we shall see in Chapter 7, every case has to be evaluated on its merits – and the evaluation is likely to be controversial.

Moving people or goods requires not only a suitable motor to move them, but a source of energy that can be delivered to the motor while it is moving. If the motor burns fuel, say to propel a car or an aircraft, the car or aircraft must be able to carry along with it enough stored energy to get it to its destination. Liquid fuels like petrol and jet fuel pack an astonishing amount of usable chemical energy into a modest mass and volume. When you squeeze the lever to pump petrol into the tank of your car, you are delivering energy at a rate equivalent to that from a small power station – perhaps 30 megawatts.

Engineers have tried for decades to devise a way to store a comparable amount of electrical energy in a comparably portable form, using a battery. A car driven by an electric motor would be much quieter and cleaner than one burning petrol or diesel fuel. But no really satisfactory battery design has yet emerged. Electrically powered delivery vehicles like milk floats have a short operating range and low top speed; nor can you yet pull into a streetcorner 'electricity station' to top up your 'tankful' of stored electricity. Electric trams and trains take their electricity from overhead cables or electrified rails; but we are unlikely ever to see an electrically powered aircraft. Liquid fuels appear to be irreplaceable for transport. As we shall see, however, they raise serious questions for long-term global energy planning. How long will liquid fuels be available, and adequately cheap? If those based on fossil fuels are available and adequately cheap, how will the resulting carbon dioxide emissions affect global warming?

Common industrial processes require physical functions similar to those of households, in particular increased temperatures and mechanical work. Industry, however, requires these services on a much larger scale. A steel-making furnace must keep many cubic metres of material far hotter than a domestic cooker. A rolling mill forming the fresh steel into usable sheets must exert forces far greater, and over a much greater area, than those used on a domestic workbench. Nevertheless the same basic physical principles of energy-use apply. Furthermore, traditional processes in major industries have long been extravagantly wasteful of both energy and materials. Modern processes are both much more efficient and much more economic, as we shall see in the next chapter.

One distinctive category of industrial process, with no real household parallel, is based on so-called 'electrochemistry'. Aluminium metal, for instance, is produced by passing powerful electric currents through cells filled with a molten compound made from aluminium ore. Electrochemistry is in itself a comparatively efficient technology; but using it on a industrial scale requires a steady supply of very cheap electricity. Because of the way it is manufactured, aluminium has been called 'solid electricity'.

In the households and offices of the 1990s, much the fastest growing type of energy use is through electronics, in telephones, radios, televisions, stereos, videos, CD players and computers of every kind. Electronics now constitutes a whole new category of energy systems, sometimes called 'information technology'. Unlike the energy uses described above, electronic technologies convert surprisingly modest amounts of energy; but they do so with remarkable subtlety and sophistication. Moreover, they usually operate essentially at ambient temperature, and accordingly lose very little energy as waste heat. Such 'ambient temperature' technologies are inherently more efficient, making functional use of a high proportion of the energy supplied. But the energy is supplied as electricity; and, as we have already noted, most traditional electricity-generating technologies are already less than 35 per cent efficient. To take full advantage of high-efficiency electronics, we should also ensure that the electricity used is generated as efficiently as possible. We shall explore the possibilities in Chapter 8.

INVESTIGATING ENERGY USE

From the mid-1970s, starting from this basic understanding of the particular physical functions that people use energy for, many researchers – at first mainly academics and environmental groups – began compiling detailed data on what people actually did with the energy they used. They quickly established that in industrial society the largest single application of fuels and electricity – anything from 25 to more than 50 per cent of the total used, in various places – is to provide low-temperature heat, to raise temperatures a few tens of degrees for space and water heating. As noted above, this is the least physically demanding use of energy. Moreover you can make it even easier simply by reducing the loss of heat from whatever you want to heat up, be it a kettle, a domestic water-tank, an industrial process-vessel or a building.

In most industrial countries the next largest application of fuels –

from 25 to 35 per cent of the total – is to supply concentrated portable energy for vehicles, especially petrol for cars. Even a casual observer, however, notices wide discrepancies in the performance of vehicles; for instance, cars in Europe travel on average much greater distances than cars in the US, for the same amount of petrol.

Countries and regions with major heavy industries like steel-making, pulp and paper production and cement manufacturing need significant amounts of high-temperature heat to raise temperatures hundreds of degrees for industrial processes. Here too, however, traditional processes allow unnecessary loss of heat. Other industries, like manufacturing consumer electronics, need much less energy per unit of output – and the products can be sold at much higher prices, tonne for tonne. Aggregating all industrial activities in the traditional way obscures the vast differences in energy-use in different industries. Focusing on specific industries and specific industrial processes reveals abundant opportunities for improvement.

Applications that can use energy only in the form of electricity – lighting, electric motors of every size, electrochemistry and electronics – constitute a surprisingly small proportion of end uses in industrial society – perhaps 8 to 15 per cent. Investigations reveal that most of the electricity used in most places is supplying low-temperature heat for space and water heating – a low-grade use for such high-grade energy. Amory Lovins refers to this sort of mismatch as 'cutting butter with a chainsaw'.

In the next chapter we shall explore all these findings, and what they imply. Must we continue cutting butter with a chainsaw? Can we – should we – change the way the world works?

7

MORE FOR LESS

You are, let us say, an Inuit. Europeans call you an Eskimo, but you know better. You have reached your destination and built yourself a compact dwelling out of thick blocks of packed snow. The snow-house provides shelter from the Arctic wind, and its thick walls of finely-packed snow crystals conduct heat only very slowly. Once you are inside, with the small doorway carefully sealed with fur-covered skins, the heat-output from your body gradually raises the temperature around you. However, this particular Arctic night is cold even by your stoical standards; the difference in temperature between inside and outside bleeds away your heat-energy uncomfortably rapidly. You could build an extra layer of snow-blocks to make the house-walls even thicker, to reduce the rate of loss of heat. But you are exhausted and sleepy. If you go to sleep in this cold, you may not wake up again. Of course you cannot build a fire; here in the tundra you are far north of the tree line, and in any case, the high flame-temperature of a fire inside a snow-house would swiftly soften the roof and bring it down on your head. What do you do? Simple: you invite one or more of your warm-blooded sled dogs to join you inside the snow-house. Each dog will contribute perhaps another 100 watts of heat-input, at a body temperature close to your own – about 35 degrees Celsius, which will not endanger the roof. Of course Inuit did not traditionally describe temperature in degrees Celsius. Instead, an Inuit would refer to a really severe outdoor temperature as a 'three-dog-night'.

The Inuit had two options: to reduce the loss of energy, or to supply more energy. The traditional approach to energy in industrial society has been the latter: to deal with a three-dog night by bringing in the dogs – that is, to compensate for the shortcomings of buildings and other energy hardware by using more fuels and electricity. Bringing in the dogs, however, has its own drawbacks; not everyone would fancy spending an Arctic night penned up with a group of powerful and aggressive energy-suppliers like Inuit huskies, notorious for fierce battles among themselves, and an equally fierce appetite for your personal provisions. Given the opportunity, an Inuit would much prefer to reduce energy losses from the snow-house, to avoid having to bring in any more dogs than absolutely necessary.

In modern industrial society, we now know that we have abundant opportunities to improve our energy hardware, and minimize the need to bring in the dogs – that is, minimize the need for fuels and electricity. The three-dog night of 1973–74 prompted energy analysts to look more closely at society's snow-house. Instead of bringing in more dogs, might it not be easier, cheaper and safer to improve the snow-house? Instead of expanding supplies of energy carriers, might it not be easier, cheaper and safer to improve the energy hardware itself?

ENERGY EFFICIENCY

Planners concerned primarily with fuels and electricity advocated 'conservation', and rapidly made it appear an option appealing only to masochists. Other analysts, however, examined entire sequences of energy use, studying their 'efficiency'. Efficiency, narrowly defined, is a concept long familiar to physicists. In general, the efficiency of an energy process is the 'useful' energy delivered, stated as a fraction or percentage of the original energy supplied. For example, if a power station converts 35 per cent of the energy in fuel into useful electrical energy, the conversion process is 35 per cent efficient – and by inference 65 per cent wasteful, if 65 per cent of the original energy is discharged as useless. A study by American physicists Marc Ross and Robert Williams, published by the American Physical Society in 1975, examined the ultimate physical limitations on the efficiencies of many familiar energy-conversion processes in households and industry. They found that the energy hardware actually in use rarely achieved efficiencies even as high as 1 per cent of the ultimate potential.

Unlike 'energy conservation', with its uncomfortable negative hairshirt connotations, 'energy efficiency' had an upbeat, positive ring to it – even in subsequent years, as it became more a catchphrase than a clearly defined physical concept. How, for instance, do you measure the 'efficiency' of a well-designed house, that keeps you warm and comfortable while requiring little or no gas or electricity for space heating? In the past two decades houses like this have sprung up all over the world, as we shall describe in more detail below. Such a house disconnects the indoors from the outdoors almost completely, no matter what the weather. It converts ambient energy from the sun, warm-blooded occupants and appliances to maintain the indoor temperature desired; but the original 'input energy' – sunlight through the windows, and heat given off by people and refrigerators and the like – is neither measured nor paid for. The concept of 'efficiency' simply does not apply. To be sure, the owner pays for the house; but the house itself delivers the energy service required, with minimal need for supplementary energy carriers. An Inuit would not have to bring in the dogs.

From the early 1970s onwards, energy analysts around the world built up a vast body of data on every variety of energy service, and the hardware available to provide it. They catalogued the data according to the specific physical nature of the service desired: temperature control over tens of degrees; temperature control over hundreds or thousands of degrees; illumination; mechanical work to move levers or turn shafts; mobility; electronics; and so on.

Studying this expanding array of data enabled analysts to identify the best technical options already available, and pinpoint the most promising directions for further development. The data revealed many specific opportunities to make dramatic improvements in efficiency, and to get 'more for less': better energy services at lower cost. Enlightened designers and planners began to act accordingly; and the traditional approach to energy began to give way to the innovative alternative. In the 1990s the energy alternative is no mere paper hypothesis. Innovative energy systems of all kinds – buildings; lighting, heating, ventilation and air conditioning; appliances; motors; industrial plant; cars – are already in service in many places, demonstrating their advantages.

INDOOR COMFORT

Consider, for instance, the largest single category of energy service in modern industrial society: providing a comfortable indoor environment in homes. Between 60 and 80 per cent of the fuel and electricity used in residential buildings is for space heating alone. Improving the building and the heating system can reduce the need for fuel or electricity to a modest fraction of this traditional level. In temperate or northern climates, where the outdoor temperature will be lower than that indoors for most of the year, you first need to reduce the loss of heat from inside the house. Warm air leaks out of a building through cracks in the structure like ill-fitting windows and doors. Heat flows out through roofs, walls, window-glass and floors. Some common building-materials – for example brick, stone, tile and glass – conduct heat entirely too well; place your hand against a single windowpane and feel the heat ebb away through the glass.

To retain heat effectively, a building should be as air-tight as possible. Its structure – the 'building shell' – should be sheathed in a layer of material that does not conduct heat well: 'insulation'. The word 'insulation' comes from the Latin 'insula', meaning 'island': insulation creates a warm island in the middle of an expanse of colder surroundings – or, in the tropics, a cool island in the heat. In general, the thicker the insulation, the slower the heat flow through it. Surprisingly enough, one of the best insulating materials is ordinary air. Trapped between close-fitting panes of glass, or in myriad tiny bubbles in plastic foam, or even in a gap or 'cavity' between inner and outer skins of a wall, motionless air reduces heat-flow substantially.

'Superinsulated' houses in Sweden, Canada, the US and elsewhere retain heat so well, even in severe winters, that they use less than half as much fuel or electricity than the average for conventional houses in the same locality, sometimes much less. For instance, the Northern Energy Home design now being sold in New England incorporates not only extra insulation but triple or quadruple glazing, indoor shutters to cut heat losses through windows at night, and air-tight construction, with a ventilation system that recovers heat from expelled air to warm the incoming fresh air. This design uses less than one-tenth the US national average amount of fuel or electricity for space heating for single-family dwellings. Moreover, it does so in a demanding climate. Because Sweden's climate is harsher still, Swedish houses have long performed thermally much better on average than US houses. Even so, the latest Swedish designs reduce the need for fuel and electricity for space-heating to levels as low as the best achieved in the US. Houses with similar high performance are springing up throughout Canada and Northern Europe.

Examples like these demonstrate the potential for improved thermal performance in new housing. To be sure, a house that performs better may cost more; but evidence indicates that the owner swiftly recuperates this extra capital cost by savings on fuel or electricity. Moreover, reducing requirements for supplementary space heating means that the heating system can be smaller and less expensive, further offsetting the extra cost of the house itself. A house which is thermally 'tight' – well insulated from the outdoors – is inherently more comfortable; indoor temperatures are more stable, minimizing draughts. If the house itself performs well enough, it does not need costly central heating; small gas or electric heaters for occasional topping up on the coldest days suffice. A really cold climate may nevertheless necessitate significant supplementary heating. A 'heat pump', like a refrigerator turned inside out, can take heat-energy from the cold outdoors and deliver it into the warm interior of the house. Using electricity to drive a heat pump can deliver almost three times as much useful heat into the house as using the same amount of electricity in a traditional electric heater.

Houses are built to last; and the great majority of houses in industrial countries were built long before the alternative approach to energy came on the scene. Altering the structure of an existing house to improve its thermal performance is both more difficult and more expensive than building it well in the first place. Nevertheless, space heating in existing dwellings uses a substantial fraction of total fuel and electricity in industrial society; and most of these houses will still be occupied for decades to come. Fortunately, even houses built to traditional standards can be modified to improve their otherwise mediocre thermal performance.

Of course traditional houses were built that way on purpose. Architects and builders have always, for understandable reasons, been essentially conservative. Their work has to last for decades, if not centuries; they tend to choose well-established materials and structural configurations, and to follow well-established customs and assumptions – partly so that they themselves know how their buildings will behave in the long term, and partly because buyers, too, tend to look for familiar features. Moreover, house building has long been governed by assumptions about the initial capital cost of the house – not about its subsequent running costs. You would have to be very rich or very foolish to buy, say, a car without thinking about how much it might cost to run. But prospective home-owners – not to mention those who finance house-purchases, by providing mortgages and other assistance – have rarely regarded the cost of running the house as a significant factor. Such a casual approach was defensible when fuel and electricity were cheap; now, however, it may be seriously misleading. The monthly heating bill is now often comparable to the monthly mortgage payment. A modest investment in improved thermal performance even in an existing house may pay for itself in only a few years, by reducing expenditure on fuel or electricity for heating.

Many North American gas and electricity suppliers now offer 'energy audits': expert technicians will inspect your house and suggest how to tighten it up. Some organizations provide more sophisticated help. In the eastern US, for instance, Princeton University has teamed up with gas suppliers in a concept they call the 'house doctor'. Two technicians with portable instruments spend one day inspecting a house, to identify leaks and other thermal flaws. They pump air into the house like inflating a balloon; the higher air-pressure inside the house makes it leak faster, and they can detect the leaks within a few minutes. An infrared viewer reveals hot spots where heat is escaping too fast through walls or ceilings. The technicians can often correct the flaws in the house immediately, with inexpensive materials. When 'house doctors' made experimental home visits in New Jersey and New York, one-day housecalls by two 'doctors' reduced gas-heating needs by an average of 19 per cent; further conventional modifications reduced gas-use by 30 per cent, for an average investment of some US$1300. At prevailing gas prices, this represented a return on the investment of nearly 20 per cent per year. With more elaborate modifications, even a house considered 'thermally tight' by US standards can reduce its requirement for energy carriers for space heating by two-thirds.

But wait a minute. Surely, by offering an 'energy audit' or a 'house doctor', a gas or electricity supplier is acting against its own interest, helping its customers to improve their efficiency – and therefore use less gas or electricity. Is this any way to run a business? It is, if the business is a regulated monopoly supplier of energy carriers – and if the regulators make the appropriate rules. In the US, for instance, the rates charged for gas and electricity are determined by a state authority called a 'public utilities commission' or some similar name. The commission hears submissions not only from the utility but from other interested parties, notably consumer and environmental groups, and sets rates accordingly. In New England, California, New York and increasingly elsewhere in the US, the commissions in recent years have been adopting an approach called 'least-cost planning', of which an early advocate was the energy innovator Amory Lovins.

The key to least-cost planning is to compare the cost of investing in new supply facilities with that of investing in improved end-use efficiency, and to select the less costly option – which now almost invariably proves to be improving efficiency. The utility goes to its customers and offers to install better insulation, more efficient lighting or appliances, or the like. The rate-making commission credits the utility with the relevant investment, and sets a rate allowing the utility a suitable return on such efficiency investments as well as on supply investments. Customers get the comfort, illumination and other energy services they desire, with lower gas and electricity bills; and the utility gets a better and more rapid return on efficiency investments than it would on long-term investment in new supply capacity.

When the concept of least-cost planning was first proposed, it met with almost universal hostility from utilities; it is still by no means generally accepted, even in the US. Utilities considered themselves to be in business to sell gas and electricity – the more the better. In the 1970s and early 1980s, US environmental organizations like the Conservation Law Foundation and the Environmental Defense Fund had to file lawsuits to force utilities to consider the potential for investing in improved efficiency instead of expanded supply. Regulators, too, began to look thoughtfully at this novel possibility. More recently, many of the utilities formerly stubbornly opposed to such a cockeyed notion have now embraced it, as eminently sensible and indeed good business.

In September 1989, for example, New England Electric joined forces with its old adversary, the Conservation Law Foundation, in a US$65 million programme to promote end-use efficiency. The utility offers to pay commercial developers the extra cost of installing the most energy efficient equipment, and install efficient lighting, cooling and heating in existing commercial and industrial buildings. The progamme should save customers more than US$100 million a year. The regulators are allowing New England Electric to recover about 20 per cent of the measured savings, a better return than the utility could make from investment in additional generating plant.

British building regulations, amended in April 1990, now set standards for maximum acceptable heat-loss through the building shell equivalent to those in force in Sweden fifty years ago. As a result, house-builders routinely construct houses whose structures are as cheap as possible, with thermal performance to match. This keeps down the initial cost of the house; the consequent high running-cost for fuels and electricity falls not on the builder but on the occupants. Moreover, all too often these occupants do not own the house; like the builder, the landlord – perhaps a local government – has no incentive to invest any more than absolutely necessary in the building structure, since the tenants have to pay the heating-bills.

Warm air rises, as colder denser air pushes in beneath it; and in far too many British houses the warm air barely slows down before it has carried valuable heat back outdoors again, leaving the colder air to soak up yet more expensive fuel-energy before it in turn departs upward. Accordingly, the two measures of highest priority in a typical British house are basic: to plug cracks and gaps in the shell of the house, to keep the colder air out; and to place a thick layer of insulation inside the roof, perhaps on the floor of the loft or attic. Thereafter other measures – extra insulation of walls and floors, double doors, double-glazing – become appropriate. The technical opportunities to improve the average thermal performance of British housing dramatically are abundant and well-documented; organizations like the Association for Conservation of Energy and the government's own Building Research Establishment have stressed the potential still unrealized.

In Britain, however, neither the government nor the suppliers of fuel and electricity have taken the opportunities very seriously, limiting their initiatives largely to rhetoric and exhortation. In the mid-1970s, environmental groups like Friends of the Earth began offering volunteer services, to insulate the houses of pensioners and low-income families; for several years from the late 1970s the British government made grants available to foster these programmes, and encourage other householders to install insulation. By the late 1980s, however, even this modest financial support had ended. Each British winter, many elderly people die of 'hypothermia' – in less grandiose terminology, they freeze to death – because their draughty flats cannot keep them adequately warm. In response, the British government for some years offered a small 'fuel supplement' to help them pay their gas and electricity bills. Activist campaigners have insisted for years that using such money to block draughts and insulate the flats – a once-for-all investment in comfort – would make more sense, instead of merely passing the money briefly through the hands of the sufferers each winter into the coffers of the fuel and electricity suppliers. However, the British government and Britain's fuel and electricity suppliers continue to regard indoor comfort as a problem of supplying energy carriers, not of improving the dwellings that use them.

The disparity between British, US and Swedish approaches to the provision of indoor comfort illustrates a crucial point, one that is all too often overlooked. Countries, states, cities and other social groups establish their own ground-rules for every human activity, including using energy. Sweden lays down building regulations stipulating high thermal performance. Britain does not. Many US states require utilities to help their customers improve the efficiency of their premises. Britain does not; and so on. Some ground-rules are formal and explicit – laws, standards, regulations, and the like. Others are simply a habit of mind, a way of thinking, taken for granted. Although these ground-rules may be inconspicuous, they are all-pervasive. They exert a powerful influence on specific choices and decisions.

No energy option, however technically promising and economically attractive, will succeed if the ground-rules are loaded against it. Accordingly, we cannot hope to change the way we use energy merely by changing technologies. Again and again, as we shall see in the coming pages, we shall have to change the ground-rules; and that will almost certainly upset somebody – usually somebody powerful. Energy policy is thus quintessentially political, in the best sense of that much-abused word. Energy policy implies energy politics – never more so than today, as the alternatives and options ramify, and the choices carry daunting global implications. Even such a fundamental amenity as comfortable shelter entails allocating effort and resources between competing alternatives: better buildings, or the same old buildings and ever more energy carriers? As the pressures intensify, the competition can only grow fiercer.

DOMESTIC APPLIANCES

After space heating, the three residential services that use most fuel and electricity in industrial society are water-heating, refrigeration and lighting. Refrigeration uses perhaps 25 per cent of electricity in particular, and lighting typically about half this much. Whereas the service-life of a house may be many decades if not centuries, the service life of domestic appliances – water-heaters, refrigerators, freezers, lights and so on – is likely to be much shorter. The resulting turnover of stock, as users replace worn-out appliances, allows them to upgrade to more efficient ones; indeed in some cases the higher efficiency available makes replacing an appliance economic even before the older model has worn out.

Heating water for washing, baths and other purposes uses from 10 to 35 per cent of the fuel and electricity supplied to a typical household in an industrial country. Like space-heating, water-heating is physically simple; but the higher temperature of hot water accelerates heat-losses from the tank and the pipes. Analysis indicates that the more you insulate a domestic hot water tank the better. Gas-fired condensing water-heaters, now coming on the market, recover the heat usually lost up the flue in water-vapour from the burning gas. Heat-pump water-heaters use only one third the electricity of conventional electric water-heaters. Coupled with well-insulated tanks they cut wastage and costs alike to a fraction of those traditionally accepted.

Of all the energy-using appliances in a typical house, the refrigerator and deep-freeze may be the least conspicuous and easiest to upgrade. Unlike lights, television, vacuum cleaners and such, refrigerating hardware operates 24 hours a day, pumping heat out of an insulated box to keep the temperature inside as low as desired. To keep heat from re-entering, the latest designs employ much more effective insulation around the body, and therefore use much less electricity. The most efficient refrigerator-freezers now on the market in the US, Europe and Japan use less than half as much electricity as traditional US designs, and those now coming onto the market less than a quarter. What is more, a higher-efficiency refrigerator may well cost not more but less; the cost of extra insulation may be more than offset by the savings on a smaller heat-pump circuit. Recent studies show no correlation whatever between cost and efficiency for refrigerators.

An informed buyer choosing a new refrigerator can therefore save both initial expense and running cost. Getting the relevant information may be the main problem. The US and many European countries stipulate that new appliances must be labelled with their operating efficiency, on a standard basis for comparison. In Britain, however, no such label is required; nor in general will the salesperson be able to tell you. Consumer and environmental groups have been lobbying for years to persuade the British government to legislate for efficiency-labelling on appliances. As this is written the government continues to reject the idea. Its impact might be considerable; one study found that if the existing US stock of refrigerators and freezers were replaced by the most efficient models on the market even in 1982, US electricity use would fall by 18 gigawatts – that is, the output of 18 of the largest power stations.

Residential lighting is less significant than water-heating and cooling, mainly because in general people turn on the lights in their homes only during the evening. Domestic lighting is still largely provided by descendants of the incandescent bulb invented by Edison and Swan more than a century ago. The incandescent bulb, however, is only about one-fifth as efficient as the fluorescent lamp. Until recently, people have preferred incandescent light to fluorescent light for domestic interiors; classical fluorescent light is stark and garish, reminiscent of the arc-lighting that preceded the advent of the incandescent bulb. Now, however, modern fluorescent fittings can deliver a light as gentle as incandescent, and indeed can be plugged into the same sockets, lighting immediately when the switch clicks.

Although an individual compact fluorescent lamp currently costs far more than an incandescent bulb, the fluorescent lamp lasts so long it is more like an investment than a short-term running cost – like buying an entire lamp, not just a lightbulb. That, however, raises another awkward question. Major lighting manufacturers have vast production capacity for incandescent light bulbs, which burn out and need replacing every few weeks. Why should these same manufacturers promote their high-efficiency long-life lamps? Tooling up for mass-production of compact fluorescents would lower their cost and make them much more economically competitive; but that would jeopardize the same manufacturer's traditional business. Conflicts like this surface repeatedly throughout the energy issue; they are not easy to resolve.

Be that as it may, by the end of the 1980s the World Resources Institute report 'Energy for a Sustainable World' could offer a startling finding. An average US household with all modern amenity services – space-heating, air-conditioning, hot water, refrigerator, freezer, cooker, lighting and other appliances – used electricity at a rate of about 360 watts per person and fuel at about 1140 watts: 1500 watts of fuel and electricity overall. An equivalent all-electric US household, getting the same services by using the most efficient end-use technologies available even in 1982/3, would use only 328 watts per person – less than a quarter as much. A similar comparison between average and most efficient Swedish households showed a still greater decrease – from 1242 to only 266 watts, less than a fifth as much. As the report noted, 'While these hypothetical households enjoy a much higher level of amenity than average households today, they would use only about 300 watts per capita, which is not only much less than present levels of household energy use in the United States and Sweden but also far less than the 1100 and 1400 watts per capita used in most Northern European countries at present to support much lower levels of amenities.'

DESIGN AND ENERGY USE: AT WORK

Think before you answer this. Which would you expect to use more fuel and electricity for space heating, air conditioning and lighting per square metre of floor: a house occupied 24 hours a day, or an office block occupied only from nine to five? One clue: remember that space heating and air-conditioning compensate for the difference in temperature between outdoors and in. If the indoor volume of a building is large compared to its outer surface-area, heat-flow outwards or inwards has less effect on the average temperature indoors. Another clue: remember that the bodies of warm-blooded occupants contribute over 100 watts of heat each, and office hardware like photocopiers, fax machines and computers considerably more. Yet another clue: remember that nine to five is daytime. Thinking rationally, would you expect the office block to need more fuel and electricity for space-heating, air-conditioning and lighting per square metre of floor than the house?

According to the US Energy Information Administration, commercial buildings in the US – including to be sure not only office blocks but also stores, warehouses, schools, hotels, theatres, and churches – use on average more than 50 per cent more fuel and electricity for space heating per square metre of floor than residences. They use nearly 550 per cent more – that is, five and a half times – for air conditioning; and an astonishing 580 per cent – nearly six times – for lighting. Why should this be so? The shell of a commercial building encloses a larger volume of interior than the shell of a house; heat losses or gains through the shell ought to have less effect on indoor temperatures. As for lighting, commercial buildings are occupied mainly during the daytime, when the sun is shining; and they tend to use fluorescent lights – five times as efficient as domestic incandescent lights. Why then should a commercial building use nearly six times as much electricity as a house for lighting a given area? Something about the energy arrangements for commercial buildings must be very strange; and this indeed proves to be the case.

Commercial buildings last a long time; and the great majority of those still functioning were built before the 1973 oil-price rise. Individual commercial buildings of course differ widely in their structures and in their energy systems; but far too many still function as though petroleum cost less than three dollars a barrel, and other energy carriers the equivalent. Consider, say, Kansas City Chambers, a hypothetical 'typical' commercial building in a typical US climate, representing the characteristics of many. Bizarre though it seems, Kansas City Chambers requires space-heating in midsummer and air-conditioning in midwinter.

Kansas City Chambers accumulates heat-energy from its occupants, from energy hardware like lights, computers and the like, and from sunlight through its windows. This heat energy is not, however, evenly distributed through the building volume; nor can it escape easily through the walls and roof, because of internal walls and floors, and because the outer surface of the building is small compared to its inner volume. To cope with this problem, Kansas City Chambers has a ventilating system running continuously, in which fuel-energy generates two air-streams, one hot and one cold. These streams are blended in appropriate proportions to deliver the desired temperature to each separate part of the building. In midsummer, however, the chilled air-stream makes rooms with comparatively little 'free' heat from occupants or energy hardware too cold; hot air must be mixed in. Conversely, in midwinter the hot airstream makes rooms with too much 'free' heat too hot; chilled air must be mixed in.

The lighting in Kansas City Chambers is equally extravagant. Although the building uses fluorescent lights with their high electrical efficiency, the bulbs are recessed into high ceilings far from work-surfaces, and behind 'diffusers' that attenuate the light; only 20 per cent of the light actually reaches the work-surfaces. Furthermore, entire offices, corridors and storerooms are lighted uniformly, to an intensity suitable for impromptu brain surgery. The lights cannot be turned on or off individually; and most of them remain on day and night, 365 days a year.

Kansas City Chambers may be a horrible example of how not to use energy in a commercial building; but it typifies practices still prevalent throughout the industrial world. To compensate for the inadequate energy performance of the building structure, the building operators install more energy hardware, like heating and ventilating systems using fuel or electricity; but the hardware itself delivers an energy performance far below optimum. After 1973, to be sure, architects and designers recognized the pitfalls of relying too heavily on energy carriers; commercial buildings erected since 1973 use only about half as much fuel and electricity per square metre of floor as the pre-1973 average. Even these newer buildings, however, can be improved.

'Energy for a Sustainable World' describes a revealing exercise carried out to lay the basis for US federal Building Energy Performance Standards. From a list of 1661 commercial buildings representing US design practices a few years after the 1973 oil embargo, the organizers chose a random sample of 168. They contacted the original design teams, and asked each to redesign the same building with improved energy efficiency. The design teams were not to violate any requirements of their original clients, nor change the overall first costs of the building significantly; and they were to use only 'off the shelf' technologies. Since few of the original design teams were expert in energy efficiency, the Research Corporation of the American Institute of Architects provided three days of training, and a workbook with practical design ideas and simplified analytical tools. On average, the redesigned buildings, costing very little more, used only about 60 per cent as much fuel and electricity as the originals.

Even pre-1973 commercial buildings, however, offer ample opportunities to improve their energy performance – often with minimal investment, merely by adjusting the controls on heating, ventilation and lighting to minimize superfluous uses. Microelectronic sensors can read temperatures, lighting levels and other relevant information at many points throughout a building, in 'real time': moment by moment. Occupants move about and switch energy hardware on and off; the sun goes behind a cloud and emerges again a few minutes later. The sensors tell a computer what is happening; and the computer adjusts the heating, ventilation and lighting accordingly. Such 'real-time control' lets the operators of a building take advantage of changing local circumstances throughout the building; they can then provide suitable environments for all the different people doing different things in different parts of the building, using fuel and electricity with subtle precision. 'Real-time control' is rapidly becoming standard on new commercial buildings; but it can also be installed in much older structures as well.

A striking example of the potential available is the New York office of a leading US environmental organization, the Natural Resources Defense Council (NRDC). NRDC has been embroiled in energy controversy for some two decades. Planning new premises in New York, they decided to turn a typical office building into a showplace for energy innovation. The existing heating and insulation were so bad that the building was 'like a space heater for the entire neighbourhood', according to NRDC. The roof, walls and windows are now super-insulated. Natural light enters from skylights and corridor windows, and through internal windows from the outer offices into the corridors. The lights are high-efficiency fluorescents; but the ceiling lights provide only background. Task lights on each desk deliver light where NRDC staff actually want it. Traditional office lighting uses about 30 watts per square metre, the NRDC office only about 5; but the NRDC office lighting is more congenial and comfortable for the staff. An occupancy sensor in each room switches the lights on when the first person enters, and off about ten minutes after the last person leaves. The high-efficiency lights give off so little heat that smaller air conditioners suffice. Each of the four NRDC floors has a separate air conditioner, so that cooling the topmost floor in midsummer, for instance, does not affect the floors below.

The energy innovations added an estimated US$500 000 to the cost of renovating the building. But they will cut fuel and electricity costs in half, saving NRDC about US$85 000 in the first year, and paying back the cost in six years – after which they will pay for themselves many times over. Since the office opened, in April 1989, it has become an even more successful showplace than NRDC might have desired, as visitors throng to tour it. The opportunities it demonstrates can be found in countless comparable buildings throughout the industrial world.

ENERGY AND THE CAR

One single energy technology has influenced the shape of the twentieth century more than any other: the car. Even in terms of energy use, the car stands out; passenger cars alone consume well over a quarter of the oil used in industrial countries. When oil supplies were dislocated in 1973 and in 1979, many different energy services felt the impact; but the most vociferous reactions came from frustrated car-drivers, especially in the US, the most car-conscious country in the world. In the eyes of its proponents around the world, the car is as sacrosanct as India's sacred cow.

Purely as energy hardware, nevertheless, the common-or-garden car leaves a lot to be desired. The basic function of a car is to carry its occupants from A to B. Rapid-fire explosions of petrol-air mixture push pistons that turn cranks to rotate a shaft; the shaft is coupled through various gears to an axle with a rubber-tyred wheel at either end. The turning wheels roll the entire car along, and eventually from A to B. So far so good: but consider a few physical details. When the car is travelling at a constant speed on level ground, its energy of motion is constant, neither increasing nor decreasing. What, then, is becoming of the fuel-energy released by the exploding petrol? It is all being dissipated as heat, in losses from the hot engine, and in overcoming air resistance and the internal friction of the car's own moving parts, especially the 'power train' of shafts and gears and the rolling tyres. The only time the petrol-energy is actually being converted into useful energy of the car is when the car is speeding up or climbing a hill.

You can describe a car's fuel 'efficiency' simply by how far it will go on a given amount of petrol – miles per gallon or kilometres per litre; or by how much petrol it requires to travel a given distance – now usually litres per 100 kilometres. For more precise comparisons you must specify a detailed 'driving cycle' – stops and starts, slow running, highway cruising, hills and valleys, and other variations that affect the amount of fuel-energy converted usefully and the amount variously wasted.

No matter how you measure the 'efficiency', however, different cars use energy very differently, even when you discount the extra energy used for car radios, air conditioning and other services that do nothing to get you from A to B except make the journey more enjoyable. The classic American 'gas-guzzler' of pre-1973 vintage might travel only 11 miles to the gallon. European cars of similar vintage and carrying capacity often did more than twice as well. Since 1973 the efficiency of new US cars has doubled; but that of European and Japanese cars remains much higher on average. Moreover, the potential for further improvement is substantial. Commercial models from major car manufacturers including Honda, Volkswagen and Chevrolet can do better than 50 highway miles to the gallon; and Volkswagen, Renault and Volvo have built prototypes that can do upwards of 80 miles.

A high-efficiency car may have an advanced engine that controls combustion electronically. It may have a 'continuously variable transmission', to optimize coupling the engine to the wheels, to reduce losses. Its tyres will have reduced rolling resistance, and its body, of strong lightweight materials, will be streamlined to minimize air resistance. According to Bill Chandler of the international research organization Battelle, these measures would double the fuel economy of an American car for only about US$500 per car; even at low US petrol prices the extra investment would pay off at the petrol pump within a year or two.

The technical and economic feasibility of more efficient cars is, however, only part of the story. To run a car you have to pay not only for fuel but for servicing, parking, taxes and other costs to which fuel economy is irrelevant. Above a certain efficiency, these other costs become predominant, blunting any interest in further improvement of fuel economy. Worse still, as world oil prices have fallen, car-owners have lost interest in fuel efficiency. In the US, the average fuel economy of new cars has begun to decrease again, as a new generation of cumbersome 'gas-guzzlers' takes to the road.

The prospect is dismaying, not merely because of the car's impact on global energy use, but because the deleterious side-effects of the car loom ever larger around the world, from urban congestion through smog and acid rain to the greenhouse effect. Burning petrol in an internal combustion engine produces unburned hydrocarbons, nitrogen oxides, carbon monoxide and carbon dioxide. More and more industrial countries are imposing regulations requiring that cars be fitted with 'catalytic converters' to clean up their noxious emissions. Unfortunately, as so often happens, correcting one problem aggravates another. Catalytic convertors do indeed reduce the immediately noxious gases in car exhausts; but they also reduce overall fuel efficiency, and thereby add still more carbon dioxide to the atmosphere. According to Bill Chandler of Battelle, 'Americans with their cars produce more carbon dioxide just by driving than the average citizen of the world produces from all activities. In the US 20 per cent of all carbon dioxide comes from automobiles, and that percentage could well grow if the automobiles become less efficient'.

However, higher fuel-efficiency alone cannot solve the problems the car is creating. Urban centres from Los Angeles to Paris to Tokyo are slowly strangling in the grip of the car; and the most efficient car is still a liability when it is idling in a traffic jam. In the latter half of the twentieth century, industrial society has laid out its cities, its commerce, industry and agriculture, and even its daily routine under the influence of the car; and this layout is literally cast in concrete. The consequent problem is not one of energy-use alone, nor environment, but even more fundamental: it is a problem of the quality of life. Using fuel in the internal combustion engine, in cars and trucks on roads to move people and goods, is a fundamental premise of modern industrial society. But the premise is revealing corollaries that may become intolerable.

Recognizing this, several major cities, among them Bordeaux, Paris and – remarkably – Los Angeles have begun studying how to rescue themselves from the car. One key move will be to revitalize and upgrade public transport, including buses, trams and urban electric trains – all of which are much more efficient, not to mention cleaner, ways to use energy to move people. But the car has countless millions of dedicated adherents; and plans to de-car the world's cities will excite vehement opposition. The confrontation over the Los Angeles plan is already heated and bitter; it will grow more so before it is resolved.

In many parts of the world – North America is the classic example – the aircraft has long supplanted the train for travel between cities. In 1980 fuel alone accounted for some 30 per cent of the operating cost of US commercial airlines, even after marked improvements in fuel efficiency after 1973. In the past ten years, new and yet more efficient aircraft engines have come into service; but fuel costs still figure prominently in airline balance sheets. Meanwhile, notably in Western Europe, passenger trains have emerged again from the shadow of passenger planes. The French Train de Grand Vitesse (TGV), capable of speeds up to 300 km/h between city centres, is attracting custom from short-haul airline flights, in part because the flying itself is now often less than half the journey time, the remainder being taken up with road travel to and from outlying airports. TGV services are being rapidly expanded, with international links. Fast rail transport is much the most energy-efficient mode of interurban travel, as well being acknowledged as the most comfortable and convenient. On this front, at least, efficient energy-use appears likely to coincide increasingly with popular preferences.

By the same criteria, transport of goods by rail is both energetically and environmentally preferable to road haulage. But the deterioration of rail services in the past two decades, in favour of official support for road transport, has seen even bulk commodities like cement and gravel switched from rail to road. Reversing the trend will be difficult, not least because road transport interests now exercise enormous political influence in Europe and North America.

ENERGY IN INDUSTRY

Traditional energy analysis alludes frequently to 'industry' as a major user of energy. Since 'industry' includes everything from mining iron ore to making microchips, the label says nothing whatever about how the energy is used, or what for. Since the mid-1970s, however, analysts have gathered a revealing body of data on energy-use in every variety of industry. One point is instantly evident. Much energy-use traditionally categorized as 'industrial' in fact fulfils precisely the same physical functions as its equivalents in households and commerce: space heating, air-conditioning and lighting for factories, warehouses and offices.

Moreover, the opportunities to improve the performance of these industrial premises are often even more obvious. Factory and warehouse doors gape open even in sub-zero weather. Buildings are sheathed in single sheets of galvanized iron, aluminium or plywood, whose insulating properties are minimal. Warmed air from hard-working heaters rises to the uninhabited upper reaches of rooms with ceilings far above the heads of staff, as cold air rushes in below. Lighting fixtures, often likewise high overhead and far from actual working areas, collect dust and grime that dim their output. The remedies are as obvious as the shortcomings. Many measures indeed can be classed simply as 'good housekeeping'. Others, like better insulation, high-efficiency lighting and real-time building controls, apply to factories just as they do to office buildings.

Activities that may with more precision be called 'industrial' can be grouped according to specific physical functions as before. They include process heating and cooling; moving solids, liquids and gases; mechanical manipulation; and physical or chemical transformation. Since the mid-1970s, physicists, chemists, engineers and industrial designers have studied the 'energetics' of these physical functions: that is, how the energy involved is converted, used and wasted. They have tracked the energy behaviour of many common industrial processes – not only entire systems but also individual components like electric motors, pumps, fans, pipes, ducts, valves, tools and controls of every kind. The studies reveal enormous potential for making industry more efficient, improving products, reducing environmental impact and creating more congenial working conditions.

The opportunities cover a range far too wide to describe here; but excellent specialist sources are proliferating rapidly. One of the latest is 'The Technology Menu for Efficient End-use of Energy', prepared under the direction of Thomas Johansson, one of the four authors of 'Energy for a Sustainable World', with the backing of Vattenfall, Sweden's State Power Board. Another is 'Competitek', an industrial subscription service from Amory Lovins and his colleagues at the Rocky Mountain Institute in Colorado, which provides regular bulletins on the latest developments in more efficient technologies. A growing number of specialist magazines and newsletters, often available free to those in the field, carry reports, articles and advertisements describing new products and processes. Some, like 'Energy Management' in Britain, are published by government departments; others, like 'Energy Today', are supported by companies advertising their efficiency innovations.

Except for certain basic materials industries like those making steel or cement, the cost of fuel or electricity is usually only a modest fraction of total company costs. In many high-technology industries, for instance manufacturing consumer electronics, fuel and electricity account for at most a few per cent of total outlays. Savings on fuel and electricity, therefore, may not of themselves be sufficiently substantial in the corporate context to elicit action to improve efficiency. Leading industries are, however, always alert for chances to improve products and productivity; and enhancing efficiency frequently accomplishes these other desirable objectives as well.

Opportunities for enhanced efficiency arise, as always, in basic 'good housekeeping' of process plant; in better flow paths for materials; in heat recovery; and of course in novel technologies. In steel-making, for instance, continuous casting creates a better flow path, and requires only half as much fuel-energy, as the British steel industry has demonstrated. In making textiles or paper, minimizing cycles of wetting and drying can save from 5 to more than 30 per cent of the energy traditionally used. Radiant heat for drying paper, fast firing of pottery, new oven-designs for baking bread, and making cement with less water are just a few of the innovations that are steadily reducing the amount of fuels and electricity used in industry.

Many industrial processes require both electricity and 'process heat' in the form of steam or hot air; paper-making is a classic example. Generating both the electricity and the heat from the same fuel can more than double the efficiency of the energy-use: what Europeans call 'combined heat and power' and Americans call 'co-generation', as we shall discuss in the next chapter. Plant designers and operators who plan and integrate the supply of fuel and electricity with its use can achieve dramatic savings in running costs, while reducing the environmental impact of the manufacturing process.

Moreover, a crucial transformation of industrial energy-use is long since in process. Industrial societies around the world have already erected, for good or ill, their major infrastructure: buildings, roads, industrial plant, indeed all that vast panoply of fabricated material that makes the societies 'industrial'. As a result, these societies no longer need substantial increments of the basic materials involved: concrete, steel and other fundamental constituents of the 'built environment'. Instead, many of their industries are now manufacturing products which incorporate a much higher 'added value' per unit of energy-use: not pig-iron and cement but computers and compact-disc players. This shift in the nature of manufacturing production alters profoundly the scale and nature of industrial energy-use, and makes any aggregated description of it not only imprecise but actively misleading.

Such 'saturation' effects influence not merely the general 'built environment' but even individual households. You can watch only one television at a time, drive only one car at a time, push only one vacuum cleaner at a time. Even allowing for 'built-in obsolescence', the manufacturers' desire to ensure that products wear out and need replacement rapidly enough, the accumulated stock of energy hardware in most industrial countries is steadily approaching saturation. Further-more, replacing one television or one vacuum cleaner with another does not increase the total amount of electricity required; on the contrary, as more efficient models arrive on the market the electricity required may decrease.

To be sure, the foregoing observations apply only to industrial societies, not to the Third World, as we shall see in Chapter 9. Here too, nevertheless, energy analysts must beware, lest traditional averages, aggregates and extrapolations blind them to the significant details of what is actually happening.

GROUND-RULES FOR COMPARISON

Traditional fuel and electricity supply analysts, confronted with innovative arguments about improving energy efficiency, have historically been loftily dismissive, waving away the innovators with a paternal pat on the head and sweep of the hand. 'Yes, yes, that's all very well, but how do you know your policy would work? How do you know it would be cheaper? How do you actually ensure that people insulate their houses, that companies install more efficient motors and lights and control systems, that you would really achieve the end-use efficiencies your theories suggest?' The questions would carry more conviction if they were not posed by analysts whose own forecasts of future supply capacities, costs and timetables have been wildly wrong for two decades.

To be sure, improving end-use efficiency will not be easy. It faces many uncertainties, technical, economic and political – those indicated above and others. What matters is how these uncertainties compare with the uncertainties facing supply options. Henceforth, anyone making energy decisions and choices must demand that comparisons between supply options and end-use options be made symmetrically. What laws, standards and regulations favour expanding supply? What financial arrangements – taxes, grants, subsidies – favour expanding supply? What political influence do suppliers exercise? How likely are supply options to fall short of expectations, overrun schedules, incur excessive costs, be politically or environmentally unacceptable? How do these circumstances compare with efficiency options?

Hitherto, fuel and electricity suppliers have defined the ground-rules, to their advantage. If end-use efficiency is to fulfil the potential it promises, supply options and end-use options must be evaluated according to the same technical, economic and political criteria. Establishing this unfamiliar symmetry will be one of the most challenging tasks facing policy-makers. End-use efficiency offers a sumptuous menu of technical options, for every energy service you can think of. Improving efficiency can cost less and free resources. It can add flexibility, and enhance the security of energy services. It can reduce environmental impact. It can win time to improve supply technologies, as the next chapter will describe. It can set a crucial example for Third World development. What is more, end-use efficiencies are already improving rapidly, as people realize the opportunities waiting to be grasped. How can we accelerate this exciting transformation of the energy scene? Energy policies mean energy politics; and energy politics are on the move.

8

SHOPPING FOR ENERGY

Thomas Edison invented the electric light bulb. But when he went into business with it, did he want to sell light bulbs? No. Did he want to sell electricity? No. He wanted to sell illumination. His initial aim was to market not just an energy carrier, nor energy-converting hardware, but the final energy service – what the user actually wanted. The distinction is profound. If your business is selling light bulbs, you want to sell as many as possible. Each time a bulb burns out you can sell another; a long-lived bulb merely deprives you of a regular customer. Furthermore, if your customers install bulbs in fittings that waste half the light, you can sell twice as many bulbs to deliver the same illumination. Similarly, if your business is selling electricity, you want to sell as much as possible. The less efficient your customers' hardware, the more electricity you can sell to deliver the same energy service. Suppose, however, your business is selling illumination. The less hardware and electricity you have to use to deliver the illumination, and the less you have to maintain the system – like replacing light bulbs – the less it costs you, and the less you have to charge your customers while making yourself a reasonable profit. If you are competing with others for a share of the illumination business, the lower the price you can charge the better your prospects.

Unfortunately, Edison's original intentions were overtaken by circumstances. As already noted, public electricity supply is a natural monopoly; competition for custom soon gave way to monopoly franchises in particular areas. Electricity prices were determined under the eye of regulatory bodies like public utility commissions, whose terms of reference paid no attention to the energy service being provided. Ere long the market had broken up: one company supplied light bulbs, another electric motors, a third electricity to run them; none of the companies provided a complete energy service, and each was interested essentially in selling more of its particular product. Efficient delivery of energy services ran directly counter to company interests.

In due course the entire energy scene split up into self-interested compartments. Fuel and electricity suppliers wanted to sell more of their energy carriers. Energy hardware suppliers wanted to keep initial costs as low as possible, even if doing so meant manufacturing less efficient hardware. Purchasers of energy hardware – buildings, industrial plant, even vehicles – generally made their choices according to initial cost. They took little account of subsequent running cost, especially when fuels and electricity cost as little as they did in the 1950s and 1960s.

In the 1990s, however, Edison's original perspective has emerged again. Analysts in major corporations like Shell, British Petroleum and General Electric now study entire energy systems. They start from the energy service desired – illumination, comfortable temperature, mobility and so on; and they work backwards, through the energy hardware, to the fuel or electricity required to run it, to the processes by which the fuel or electricity are supplied. Optimizing the entire energy system, to increase efficiency, reduce cost, and minimize environmental impact, can also make the system more flexible and adaptable, improving the service it delivers. As the previous chapter indicated, energy hardware now available can deliver energy services far more efficiently than traditional assumptions have hitherto acknowledged. Furthermore, the better end-use technologies surveyed in the previous chapter can be driven by better supply technologies – not only more efficient but environmentally much less detrimental, and often cheaper.

PRODUCING 'PRIMARY ENERGY': THE HAZARDS

Modern industrial society gets almost all its fuel from underground, either by scraping away the surface to lay bare the fuel, or by going down after it, either in person or with a drill rig. So-called 'open cast' or 'strip-mining' of coal, in which enormous diggers gouge away the topsoil and overburden of rock to expose the coal seam, can create moonscapes of sterile devastation. The Appalachian coalfields in the eastern US, the gaping gashes in the Czech countryside and the vast scar at An Tai Bao in China bear appalling witness to the worst excesses of this approach.

In Federal Germany and Britain, however, open cast coal mining has long been rigidly regulated. A mining company must preserve the topsoil and rock that have been removed. After each successive strip of coal has been extracted, the company must return the rock to refill the hole, and replace the topsoil; it must then level, grade and landscape the surface, to restore as closely as possible the original topography of the terrain. The noise, dust and heavy traffic created by open cast mining mean that it will never be popular with its neighbours; but the best practice is far preferable to the traditional. It is also, however, significantly more expensive. A country that allows its open cast coal mines to ravage its landscape can sell its coal more cheaply than a country insisting on restoration. When environmental standards differ so markedly from country to country, classical economics leads to conclusions and decisions that may be profoundly unsatisfactory on a global basis.

Underground mining of coal creates similar problems. Historically, mining coal underground has been the quintessential example of a physically dangerous, dirty and brutally demanding human activity. Some countries – notably, again, Federal Germany and Britain, and more recently also the US – have passed laws laying down strict safety standards and procedures for miners working underground. Innovations in mining technology, like the heavily automated 'long-wall' technique pioneered in Britain, have dramatically reduced the sheer drudgery involved in mining coal. But the capital cost is high; moreover, the miners must be intelligent and skilled to operate the hardware, and expect to be paid accordingly. Removing coal from below ground may cause 'subsistence', as the surface above sags into the subterranean cavity. Above ground, the ugly mountains of 'pit spoil' have long disfigured the neighbourhood of deep mines, threatening 'slippages' like the hideous landslide that engulfed the Welsh village of Aberfan in 1966, killing 144 people.

Responsible mining companies now accept that spoil tips must be shaped, landscaped and planted with grass, shrubs and eventually trees; indeed some spoil tips have thus been converted into attractive hilly parkland. The world price of coal, however, is likely to be established by suppliers less scrupulous about mining safety and environmental controls. Such suppliers employ minimally skilled labour at minimal wages, working with comparatively inexpensive and primitive technology like picks and shovels; and they are indifferent to the grievous damage they inflict on their surroundings. Once again, the classical economic marketplace cannot cope with such social and environmental disparities.

Oil production and transport likewise entail impacts that may not show up in the traded price. Leaks from offshore platforms are a perennial problem, dating back to the ugly Santa Barbara spill off California in the late 1960s. Once the oil is on the surface it must be carried to a refinery, usually by pipeline or tanker-vessel. The associated environmental hazards are all too well documented, evoked by names like Torrey Canyon, Amoco Cadiz and Exxon Valdez.

Even natural gas, the 'cleanest' fossil fuel, has an impact – sometimes literally, as an explosion. Reports indicate, for example, that Soviet natural gas pipelines experience explosive leaks with alarming frequency. Such accidents underline a much more subtle hazard of natural gas. The world's natural-gas production and transport facilities are notoriously leaky – some, admittedly, much worse than others; the trans-Siberian pipeline, one of the longest, also appears to be one of the leakiest. A natural gas leak is much harder to detect than an oil leak; natural gas is not only invisible but odorless. Natural gas is almost pure methane; and methane proves to be one of the more damaging 'greenhouse' gases, some 30 times worse than carbon dioxide, molecule for molecule, although it does not last as long in the atmosphere. Natural gas has been billed as the most desirable fossil fuel, because when it is burned it produces only about 60 per cent as much carbon dioxide per unit of heat as coal. However, as natural gas production expands, unless supply facilities can be tightened up, natural gas leaking from them into the atmosphere might well offset this advantage.

Like the fossil fuels, uranium is extracted from underground; and like coal-mining, uranium mining is physically hazardous and environmentally damaging. Like coal, uranium leaves behind spoil tips that disfigure the landscape. Uranium spoil is called 'tailings': finely-divided sand containing radioactive radium that can contaminate runoff water. Hundreds of millions of tonnes of tailings are already stockpiled at uranium mines around the world. Moreover, the radium in the tailings piles gives off radioactive radon gas, now identified as a major health hazard.

Apart from the fossil fuels and uranium, one other 'primary energy supply' technology must not be overlooked. Hydroelectricity, generated by water falling through turbines, is the classical 'renewable' energy. Sunlight evaporates water from lower terrain, lifting it against gravity to deposit it higher. But hydroelectricity, as traditionally generated, entails collecting upland surface water behind a dam high enough to make the water fall a considerable distance. This usually entails flooding a significant area of land, indeed often a quite staggering area; some hydroelectric 'reservoirs' cover thousands of square kilometres. The flooded land is lost to animals, plants and people, usually very much against the will of the said people. Disturbing the water system of the locality can propagate water-borne parasites and other hazards, like the snails carrying the liver-flukes that cause a debilitating disease called 'bilharzia' around the Aswan high dam in Egypt. The sheer weight of water in a deep reservoir may even trigger an earthquake – precisely what anyone downstream of a major dam most dreads.

In sum, then, all so-called 'primary energy production' processes – coal-mining, oil and gas extraction, uranium mining and hydroelectric generation – entail major hazards, to people themselves, to their environment or both. Using the best available technologies and the best practice can reduce the hazards; but the 'best' may cost more, and put the scrupulous supplier at a disadvantage compared to less scrupulous competitors. Legislation and regulation play a key role in defining and maintaining acceptable standards; but laws and standards may not be uniform between different countries, a problem causing increasing international friction.

Inescapably, therefore, reducing the environmental hazards of 'primary energy production' means above all reducing the need for what planners call 'primary energy' – for coal, oil, natural gas, uranium and hydroelectricity. However, as the previous chapter described, this does not in any way imply cutting back on energy services: on the contrary. Indeed, just as end-use energy hardware can be much more efficient, so can supply technologies – especially those for generating the most versatile energy carrier, electricity. Moreover, efficient generating technologies can also be more acceptable environmentally.

FOSSIL FUELS: IMPROVING ON TRADITION

To generate electricity from fossil fuel – coal, oil or natural gas – you burn the fuel to release heat energy and create hot high-pressure gas. As the gas expands, it pushes a piston or spins a turbine, which turns an electricity generator. The process invariably produces not only noise but also waste gases from the burnt fuel; coal also produces solid waste. All of these side-effects are more or less environmentally troublesome; and all tend to reduce the overall efficiency of electricity generation. Modern innovations have demonstrated, however, that even the classical approach to generating electricity, by burning fuel, offers considerable opportunities to improve efficiency and diminish environmental impact.

Throughout the 1950s and 1960s engineers sought to improve efficiency simply by making power station units larger. A larger boiler or turbine has less surface area compared to its volume; it therefore loses heat less rapidly to the surroundings. Furthermore, you can double the electrical output of a unit without doubling the amount of material it contains; the capital cost per unit of capacity should be smaller. As well as increasing the size of units, engineers also boosted the 'steam conditions', designing boilers that could superheat steam to very high temperatures and pressures. The efficiency of a turbine-generator depends on the temperature drop from the turbine inlet to the final outlet – the bigger the drop the higher the efficiency. Accordingly, engineers arranged to have the steam emerge at the outlet and strike pipes carrying cold water, condensing the steam back to water at a temperature well below boiling. Such a 'condensing turbine' coupled to a generator at the highest achievable steam conditions can convert about 40 per cent of the fuel energy into electricity. That number is worth pondering. A traditional large coal- or oil-fired power station, even of the most modern design, wastes at least three-fifths of the fuel energy fed into it. At most two-fifths emerges as electricity. The remainder emerges as tepid condenser-water, too cool to be of any use.

The inherently low efficiency of traditional electricity generation has been compounded by the problem of what to do with the waste gases from the burned fuel. They include carbon dioxide, sulphur dioxide and nitrogen oxides, in proportions depending on the chemical composition of the fuel. Sulphur and nitrogen oxides – SOx and NOx – are precursors of acid rain. Most industrial countries now impose stringent limits on SOx and NOx from power stations, like those laid down by the 'Large Combustion Plant Directive' of the European Commission and similar legislation in Japan and the US.

To comply with the limit on SOx, a power station operator may buy low-sulphur coal, and pay a premium price for it. Alternatively, the operator may install a so-called 'flue gas desulphurization' or FGD plant, to remove the sulphur from the combustion gases before they go up the stack. By the end of 1988, 434 FGD plants were operating in eight different countries, including 242 in the US and 122 in Federal Germany. Unfortunately, however, an FGD plant is a huge and bulky unit that may account for one-fifth of the total capital cost of a whole power station. Depending on its design, it may require hundreds of thousands of tonnes of limestone 'sorbent' each year, with which to trap the sulphur chemically. In some FGD designs, the sulphated limestone emerges as good quality gypsum suitable for use as a building material; but the amount produced in Federal Germany alone represents far more gypsum than the country's building industry can use. Newer designs, notably one called 'Wellman-Lord', recycle the sorbent and produce commercially saleable sulphur; but they are more complex and costly, and even the world sulphur market might not be able to use the output that could readily arise.

These drawbacks are compounded by yet another. An FGD plant significantly reduces the overall efficiency of the power station; the station has to burn more fuel for each unit of electricity sent out. Using FGD to limit sulphur emission therefore aggravates carbon dioxide emission. FGD helps to mitigate acid rain, but in so doing worsens the greenhouse effect. At the moment FGD appears the preferred technology to cut down sulphur emissions from large power stations already in operation. As we shall discuss shortly, however, other options with fewer drawbacks may soon take over.

Power station operators can reduce nitrogen oxide – NOx – production from existing boilers by fitting so-called 'low-NOx burners'; many electrical utilities are doing so already. Low-NOx burners, however, can reduce NOx only to moderate levels; and some countries, notably Japan, now stipulate NOx emission levels so low that power station operators must fit so-called 'selective catalytic reduction' or SCR units onto their plants. SCR, like FGD, is costly and complicated, and lowers the overall efficiency of a power station; so SCR, like FGD, also aggravates the greenhouse effect.

Industrial society gets so much of its electricity from coal-fired power stations already operating that operators have to resort to short-term emission-control remedies like FGD, low-NOx burners and SCR. Despite their shortcomings they can be 'backfitted' to existing plants, to alleviate at least some of the associated environmental problems. In the longer term, however, trends now emerging may radically reshape electricity supply. The catalogue of technical supply options alone is expanding at a startling rate; and suppliers are revising their traditional assumptions, in ways that favour the innovative options.

Until the late 1970s, electrical utilities almost invariably assumed that a bigger power station was a better power station – cheaper and more efficient. They ordered individual boilers and turbo-generators with capacities of 900 megawatts and even larger. Unfortunately, although such enormous units promised in theory to be cheaper and more efficient, they usually proved in practice to be anything but. Theoretical savings in capital cost were wiped out by construction schedules that overran not by months but by years, sometimes extending to a decade or more. Even when large units eventually came into service they suffered much more frequent breakdowns, partly because they were simply more complicated than smaller units, with more parts to go wrong. They therefore required more backup plant, adding to the total capital cost of the system. Furthermore, planners conceded that they could not estimate with any confidence the level of electricity use a decade hence. No one could therefore guarantee that, when a large power station at last came on stream, customers would be ready to use its output and pay what it cost.

Accordingly, electricity supply planners in industrial countries like the US, Scandinavia and elsewhere have recently turned their attention away from enormous units, toward those in sizes below 400 megawatts. Smaller power stations, that can be ordered, built and brought into service in well under five years, make life much easier for supply planners. They can anticipate changing patterns of electricity use, and order small stations to match. A small station is easier to site. It needs less cooling water; it is less obtrusive and less damaging to local amenities; and it provokes much less hostility from the local community. A small station can be erected with hardware built mostly in a factory instead of on site. This enhances quality control and minimizes site labour problems, two major reasons for delays and cost overruns on large stations.

The rebirth of enthusiasm for smaller power stations substantially widens the range of technical options available. At unit sizes from, say, 50 to 250 megawatts, cleaner and more efficient technologies with names like 'combined heat and power', 'cogeneration', 'fluidized bed combustion', and 'combined cycles' come into their own.

'Combined heat and power' and 'cogeneration' are respectively the European and American designations for essentially the same concept. As both terms indicate, a 'CHP' or 'cogen' plant burns fuel to produce both useful electricity and useful heat, with the emphasis on 'useful'. In some countries – for instance Federal Germany, Sweden, and more recently the US – CHP/cogen is well established; in others like Britain it remains almost unexploited. As noted above, a conventional fossil-fired power station is designed to convert as much fuel energy as possible into electricity alone – perhaps 40 per cent at best, because the conversion process runs up against a fundamental physical limit. Conversion to electricity alone wastes at least 60 per cent of the fuel energy; it emerges mainly as warm stack gas and tepid condenser water, too cool to be of much interest. A CHP/cogen plant, by contrast, does not try to squeeze every last possible unit of electricity out of a kilogram of fuel. It can therefore produce, in addition to electricity, steam or water still hot enough to be valuable in its own right. A CHP/cogen plant can convert well over 80 per cent of the original fuel energy into a blend of electricity and useful heat.

Small CHP/cogen sysems, with an output of up to 40 megawatts or so, are often based on internal combustion engines, burning diesel fuel or petrol. The engine drives an electrical generator, much like the portable generators you sometimes see on building sites. However, instead of discharging the hot exhaust gas from the engine directly to the air, the unit passes it through a bank of pipes to heat or boil the water in them. To American engineers this is a 'heat recovery steam generator', to British engineers a 'waste heat boiler'. An onlooker might find the distinction revealing: it is 'waste heat' no longer, since the steam or hot water can now be used. At the smallest end of the range, Fiat makes a compact self-contained cogen unit it calls a 'total energy module' or Totem, based on an ordinary small car engine. A Totem unit can generate enough electricity and space heating for a sizeable dwelling.

Larger CHP/cogen units, in sizes up to more than 200 megawatts, burn coal, oil or gas to raise steam to drive a turbine-generator, just like a power station. The operator of a CHP/cogen plant, however, can tap off useful steam either partway through the turbine – a 'pass-out' turbine; or at its outlet end – a 'back-pressure' turbine. The operator can therefore adjust the balance between electricity and steam-production from a given quantity of fuel. In Sweden, Federal Germany, and elsewhere, many local municipalities own and operate CHP plants to provide 'district heating' for their communities. Steam or hot water is delivered through insulated underground pipes into all the buildings in an entire neighbourhood, for space heating and hot water. The buildings do not therefore need separate boilers or furnaces, nor separate flues or chimneys to remove combustion gases.

In a chilly February, say, the CHP plant operators expect to have to supply as much heat as possible, so the electricity output from the plant will be comparatively low. In a balmy August, however, the community will need only a modest amount of space-heating. The CHP plant operators can adjust the output, emphasizing heat or electricity as desired, and sell the surplus electricity to the grid. In Sweden, for example, small CHP plants all over the country provide local district heating, and contribute a significant fraction of the total electricity supply available nationally. Their overall fuel efficiency is twice that of fuel-burning stations that generate electricity alone.

Some CHP/cogen plants provide 'industrial cogeneration'. As indicated in the preceding chapter, a major industrial installation like a steel mill, paper mill or chemical plant requires both electricity and heat in impressive quantities, perhaps well over 100 megawatts in all. An industrial cogen plant can supply both from the same fuel, more than doubling the fuel-efficiency of the process and lowering production costs accordingly.

Using conventional boiler technology for CHP/cogen, however, raises the usual environmental problems of SOx and NOx emissions, especially since district heating and industrial cogeneration both imply siting plants in urban areas. Fitting stack-gas cleaning equipment like flue-gas desulphurization (FGD) or selective catalytic reduction (SCR) might make CHP/cogen hopelessly uneconomic. In the 1980s, however, a new combustion technology has emerged, that may gradually supplant traditional boilers and furnaces, especially for burning coal. It is called 'fluidized bed combustion', or FBC.

Unlike traditional combustion plant, FBC can reduce SOx emissions by more than 90 per cent, by trapping fuel-sulphur in solid form during the combustion process itself, with no need for stack-gas cleaning. Since FBC operates at a temperature several hundred degrees lower than conventional combustion, it produces very little NOx; simple design provisions can reduce NOx even further. With suitable fuel-handling hardware, FBC can burn almost anything combustible – low-grade coal, colliery washings, mine spoil, wood and wood waste, urban refuse, industrial waste, even sewage sludge – cleanly and efficiently. It can switch from one fuel to another, or blend different fuels as the operator requires.

FBC appears to be ideal for CHP/cogen. Within the past decade, major engineering companies like Ahlstrom of Finland, Lurgi of Federal Germany, and Foster Wheeler and Combustion Engineering of the US have constructed and commissioned more than 350 FBC units around the world, of which more than 50 are CHP/cogen units. More are being ordered by the month. By 1989 Lurgi alone had 16 FBC power plants in operation, with another 21 already on order, to come into service by 1992; Ahlstrom has more than 70 FBC units in operation or on order in a dozen countries, including a growing number of CHP/cogen units. Many of these units are in environmentally sensitive sites, where traditional coal-burning plant might now be simply unacceptable. The Ahlstrom Ace Cogeneration Plant, in California, has to meet some of the most stringent air-quality standards anywhere in the world; the Lurgi FBC unit in Duisburg, Federal Germany, is sited right next to the offices of the town council.

GAS TURBINES

CHP/cogen units fit neatly into the new philosophy of electricity supply now emerging, based on small power plants with high efficiency and minimal environmental impact. So do 'gas turbines'. A gas turbine is a compact first cousin of a steam turbine; but it can withstand much hotter gases expanding through it, at a rather different rate. Industrial gas turbines, direct descendants of aircraft jet engines, have been around since the 1940s. Major manufacturers like General Electric in the US and ASEA Brown Boveri of Sweden and Switzerland now offer gas turbine-generators with outputs of more than 200 megawatts; larger sizes are coming off the drawing-boards.

A gas turbine is comparatively cheap to build; but it usually requires premium fuel – jet fuel, light fuel oil or natural gas – and is therefore more expensive to run. Moreover, a simple gas turbine is an inefficient way to generate electricity; the exhaust gases, like those of a jet engine, carry away too much heat. Nevertheless, in the past decade gas turbines have become one of the fastest-growing electricity supply technologies, for two key reasons. One, of course, is the unexpected abundance of cheap natural gas. The other, paradoxically, follows directly from the inherent inefficiency of a gas turbine. Commercial gas turbines now on the market can accept combustion gases at a temperature of more than 1200 degrees Celsius, compared to perhaps 565 for a steam turbine. The gases emerge from the turbine exhaust at a temperature amply high enough to boil water and raise steam, in a 'heat recovery steam generator' or 'waste heat boiler', as mentioned earlier. You can then feed this steam into a steam turbine, to generate yet more electricity from the original fuel-energy. Running a high-temperature gas turbine in tandem with a lower-temperature steam turbine, to extract as much electricity as possible from the total temperature drop available, is called a 'combined cycle'. A combined cycle plant can convert upwards of 45 per cent of the original fuel energy into electricity, compared with an average of perhaps 35 per cent for a modern conventional fossil-fueled station – that is, about 25 per cent more electricity from the same amount of fuel.

Now, wherever natural gas is cheap and abundant, combined-cycle 'CC' plants have surged to the top of the table in popularity. Electricity suppliers can build them rapidly, with a 'module' of perhaps 200 megawatts coming on stream in two years or less, and add extra units step by step as necessary. Enthusiasm for this approach has burgeoned not only in the US and Britain, where natural gas is available directly through pipelines from wells, but also in Japan, burning natural gas from foreign suppliers delivered to Japan by tanker in liquid form. One of the largest CC plants thus far in operation, for example, is the Futtsu plant of Tokyo Electric Power on Tokyo Bay. It includes two 1000-megawatt blocks, each with seven coupled gas and steam turbines. In the US, the growing popularity of CC plants has prompted manufacturers like General Electric to develop gas turbines far larger than any aircraft jet engine, with ever-higher 'inlet temperatures'; the aim is to build CC plants with efficiencies of 50 per cent or even better. In Britain, in the wake of the upheaval caused by 'privatizing' the electricity supply industry, private entrepreneurs are planning to build a number of CC stations, burning natural gas from the North Sea. Some of their output will be sold to the electricity supply system and some directly to major industries like ICI and British Steel.

Some analysts express one significant reservation about CC plants generating electricity from natural gas. A sizeable fraction of the electricity used in a country like Britain delivers only low-temperature heat for space- and water-heating. For these undemanding purposes, burning gas directly at the point of use is much more efficient and environmentally acceptable overall than converting the gas to electricity, even in a CC plant. A CC plant still wastes more than half the original fuel-energy of the natural gas, and accordingly doubles the amount of carbon dioxide released in providing the space-heat or hot water.

Be that as it may, electricity suppliers would build even more natural gas CC plants, if they were not so uncertain about how much natural gas will be available in the longer term, and how much it will cost. The cost of electricity from this type of plant depends mainly on the cost of the premium fuel it burns. If the price of natural gas rises significantly – as some believe it will – natural gas CC plants may become dauntingly uneconomic long before the end of their anticipated operating life.

With this problem in mind, even in the 1940s Shell engineers were trying to marry the advantage of cheap construction and cheap operation, by developing a gas turbine that could burn ordinary inexpensive coal. Unfortunately, the hot combustion gases from conventional coal-burning are laden with dust particles fused into tiny fragments of glass by the high temperature. When the dust-laden gases strike turbine blades, the blades wear away as though being sandblasted – which indeed they are. The Shell engineers abandoned their efforts as fruitless. In the 1990s, however, the coal-fired gas turbine has at last arrived – in two quite different technologies, now competing head-on for precedence.

One is based on the versatile concept mentioned above – fluidized bed combustion, or FBC. FBC units now in service almost all operate essentially at atmospheric pressure. You can also, however, operate an FBC unit inside a pressurized shell. The hot gases from combustion then emerge at about twenty times atmospheric pressure, at perhaps 850 degrees Celsius. Although far hotter than the hottest available steam, this is well below the temperature at which coal ash melts; so the dust-particles are not glassy but soft, and much less abrasive than fused ash.

The pressurized hot gases can therefore be cleaned up and fed directly into a gas turbine, and thence to raise steam for a steam turbine – coal-fired combined cycle plant, based on pressurized fluidized bed combustion or PFBC. The world's largest engineering corporation, ASEA Brown Boveri (ABB), is now building three PFBC power stations – one in Ohio in the US, one in Spain and one two-unit plant right in the heart of Stockholm, Sweden. As might be expected, the Stockholm city authorities have stipulated very tight limits on SOx and NOx emissions from the Stockholm PFBC plant; but ABB engineers are confident their design can comply. ABB is already at work on a 330-megawatt PFBC plant for another site in Ohio.

Nor is pressurized FBC the only route to coal-fired combined cycles. Companies like Shell, Texaco and British Gas, doubtful about the long-term future of natural gas supplies, wanted to manufacture a high-quality substitute from coal. They called it 'synthetic natural gas' – another teeth-jarring contradiction in terms, like 'artificial genuine' – or simply SNG. In due course the various companies came up with a range of new 'coal gasification' technologies all of which could produce, for a price, SNG. But by that time the price was too high. Underground reserves of the real thing were proving so copious that no substitute could hope to compete. Gradually, however, a different angle came into view. Although no one appeared to want SNG, people and governments wanted cleaner and more efficient ways to generate electricity. The various coal gasification technologies all produced, after the first stage, fuel-gas of moderate quality, that then required further costly processing to make pipeline-quality SNG. Why not, instead, take the fuel-gas straight from the 'gasifier' into a gas turbine, in a combined cycle plant? Coal gasification lent itself readily to sulphur-removal technologies long familiar in the chemical industry; and combined cycle operation would boost overall efficiency significantly. By the early 1980s, a series of demonstration projects were underway, and another energy acronym sprang up in a thicket already nearly impenetrable except by the intrepid: the technology was called 'integrated gasification combined cycle', or IGCC.

One demonstration project in particular caught the imagination, not least because of its name: the 94-megawatt Cool Water IGCC plant, in the Mojave desert of California. At a time when power plants routinely overran their original costs and construction schedules, Cool Water, based on Texaco's coal-gasifier, was completed not only on time but actually under budget. From 1984 to 1989 Cool Water did everything its designers had asked of it, more reliably than they had even expected; and its SOx, NOx and particulate emissions were only a modest fraction of California's already stringent standards. The one thing it could not do was compete with low-priced natural gas for electricity generation. After its successful five-year demonstration programme the plant was mothballed, and due for demolition. In late 1989, however, it received a reprieve; Texaco repurchased the rights to use it for a startling further project: to generate combined-cycle electricity by gasifying a mixture of coal and sewage sludge.

Other IGCC plants recently commissioned include Shell's Deer Park unit at its vast site near Houston, and Dow's Plaquemine unit near Baton Rouge. In 1989 Shell won the order for the first IGCC plant in the Netherlands, at a site with the memorable name of Buggenum; if it succeeds, another and larger will follow. Electricity suppliers in a number of countries are watching closely to see which of the acronyms, IGC or PFBC – integrated gasification combined cycle or pressurized fluidized bed combustion – wins the race toward full commercial feasibility: unless, of course, both are overtaken by other supply technologies. The leading contender now is undoubtedly natural gas combined cycles. At the moment no other fossil-fuelled electricity supply technology looks as promising, economically and environmentally. But natural gas combined cycles, too, may eventually be superseded, by supply technologies not based on fossil fuels at all.

NUCLEAR HOPES REKINDLED?

A vociferous and influential body of opinion continues to insist that the key to future electricity supply is uranium – that is, nuclear power. More than 400 nuclear power reactors are already in operation in 26 countries, with about 100 more under construction – a total generating capacity of some 400 000 megawatts. Nuclear power's advocates point out that it emits no NOx, SOx, nor carbon dioxide; that makes nuclear power, they claim, at least a significant partial answer to both acid rain and the greenhouse effect. They maintain that nuclear power can meet the most demanding standards for safety. In their view, the management and disposal of radioactive waste is not a scientific nor technical problem but merely – their word – requires 'political will' to implement appropriate measures. Decommissioning of shutdown nuclear plant can be deferred for a century or more if desired, to allow radioactivity to decay and decommissioning funds to earn enough interest to cover whatever the cost may be. If uranium reserves do not keep pace with requirements, the plutonium-fueled fast breeder reactor will take over, making its own plutonium as it generates electricity.

The arguments have not changed much over the years – except in one respect: and that respect may be crucial. Until recently, nuclear power advocates also declared that nuclear power was the cheapest form of electricity. However, actual experience of nuclear operation is taking the place of hypothetical estimates and prognoses; and the economic status of nuclear power programmes around the world now looks at best questionable. In the US and Britain, the cost of nuclear electricity is conceded officially to be substantially higher than that of electricity from fossil fuels. Moreover, the financial risks associated with safety, waste disposal and decommissioning are so uncertain as to be essentially unknown. In consequence, investment analysts and financial advisors are less and less inclined to put up the capital for new nuclear plants. Partly as a result, orders for new nuclear plants have dwindled to a trickle. Global nuclear power capacity appears likely, therefore, to peak in the mid-1990s as plants already under construction are completed. Thereafter, as older plants are shut down, nuclear capacity may well enter a steep decline.

That does not mean, to be sure, that nuclear power will vanish. Quite apart from its intractable leftovers, which will still be evident for centuries, the possibility remains that nuclear power might undergo a renaissance in the twenty-first century. Major engineering corporations like Westinghouse and Mitsubishi are confident that nuclear power will be reborn, albeit possibly in a significantly different guise. To that end they are focusing on new designs of nuclear plants, better able to fulfil the criteria now considered essential. The aim is to design nuclear plants to be 'inherently safe', probably smaller and simpler. Some design teams favour a so-called 'high temperature' reactor, probably cooled by helium gas, whose core, fuel and all, is entirely ceramic, and therefore far less susceptible to meltdown, even under the most extreme accident conditions. Before nuclear power can be reborn, however, the problems of nuclear waste disposal and decommissioning must be convincingly resolved; performance and reliability must markedly improve; and capital costs must come down. Commentators outside the nuclear industry are not holding their breath.

RENEWABLE ENERGY

Anyone even slightly familiar with the history of human use of energy must find it incongruous to hear planners refer to wind and water power as 'alternative', especially when the same planners designate nuclear power as 'conventional'. Indeed, human use of energy from wind and water predates human history – written history, at any rate – in at least some parts of the world. Now, however, as using fuel creates and aggravates environmental problems, attention is turning more and more to using energy that does not come from fuel: natural ambient energy. Almost all this energy comes, as we have seen, from the sun, as high-quality solar radiation. It drives the earth's physical, chemical and biological processes, yielding up its quality until it is radiated back into space as low-grade heat. If and when the sun stops supplying this energy, we shall not be around long to miss it. Accordingly, for our human purposes its supply is always being replenished: it is 'renewable' energy.

We can intervene in or tap into natural energy processes in a variety of ways. In the preceding chapter we saw, for instance, how buildings help to reorganize natural energy flows, to give us more comfortable surroundings. We do not, however, call buildings 'energy supply' technologies, partly because we gain the benefit they provide immediately and directly, and partly because we scarcely notice that it is being delivered. 'Supply' usually refers to a measurable commodity or service, like a tonne of coal or a unit of electricity. A seafarer watching sails droop in a flat calm might wish for wind; but only a distinctly unusual seafarer would wish for a 'supply of wind energy'. In general, 'renewable energy' technologies become 'supply' technologies only when they produce an output that can be measured, and probably charged for: that is, fuel or electricity. In the parlance of most planners now, 'renewable energy' is natural ambient energy that people have converted into fuel or electricity. This fuel or electricity can then be used to run suitable hardware, to provide the energy services people want. For the purposes of this chapter, therefore, 'renewable energy technologies' are technologies that supply fuel or electricity.

Ambient energy has two potentially awkward attributes. Most of the time, in most places people can go, natural energy flows are not sufficiently concentrated to be converted rapidly into useful quantities of fuel or energy; yet sometimes, in some places, natural energy flows are entirely too concentrated for people to manage, or indeed withstand. The sunlight falling at high noon in summer on a square metre of flat land at temperate latitudes delivers perhaps 100 joules per second – that is, 100 watts: very diffuse compared to the energy released from a bonfire on the same square metre. Clouds diminish the energy flow further, to say nothing of winter and nightfall. Hurricane winds, on the other hand, can deliver an onslaught of raw natural energy as concentrated as that released from high explosive. Natural energy flows vary widely and unpredictably. To convert them reliably and efficiently into fuel or electricity is a major challenge.

BIOMASS ENERGY

Consider, first, fuel. The only process we know of that converts natural energy directly into fuel – that captures the energy flow and stores it – is 'photosynthesis'. Green plants use sunlight to convert carbon dioxide and water into energy-rich starches and sugars; as yet, however, we do not know how they do it. No matter how bright the sunlight, photosynthesis takes place too slowly for us to see the mass of a plant increase; but increase it certainly does. When the sun is shining, green leaves inhale carbon dioxide gas, and roots absorb nutrients in water solution; the plant transforms the ingredients into solid living matter: so-called 'biomass', full of concentrated stored energy. A few plants, for instance the water hyacinth, and the tropical water-lily with its raft-like leaves, grow so fast that an onlooker can just about see them getting larger. Others, of course, grow so slowly that little change is apparent even from year to year.

The terms 'biomass' and 'biomass energy' are of recent coinage: but the concept is far older even than wind and water power. That first intentional fire described in Chapter 1 was releasing biomass energy. Until the advent of fossil fuels, all fuels – firewood, charcoal, dung, and animal and vegetable fats – were 'biofuels', whose energy originated from sunlight stored by living matter. Strictly speaking, even fossil fuels are biofuels; their energy too comes from sunlight stored by living matter. In today's terminology, however, 'biofuels' are those containing sunlight energy that reached the earth at most a few centuries ago, to be stored in trees now ancient; most biofuels now of interest store energy that reached the earth during the twentieth century.

Every year the earth's growing plants store in their biomass about ten times as much energy as the earth's human population uses in the same year. A green leaf is not a very efficient 'solar collector', storing only about 1 per cent of the energy that falls on it; but there are a lot of green leaves. Perhaps 15 per cent of the energy that the people of the earth now use is biomass, mostly in the form of firewood and mostly in the developing world; see the next chapter. In the industrial world, energy once supplied by biomass is now almost entirely supplied by fossil fuels; but shifting the balance back toward biofuels looks increasingly attractive. Biomass generally contains less ash and much less sulphur than coal. Unlike fossil fuels, biomass grown and used at a renewable rate does not increase the concentration of carbon dioxide in the atmosphere; it merely returns to the atmosphere the carbon it has recently extracted. Biomass may therefore become a key factor in controlling the 'greenhouse effect' of global warming and climate change.

Biomass appears in many forms, and can be converted into a variety of highly serviceable biofuels – solid, liquid or gaseous, according to what you want to do with the fuel. Suppose, for instance, you start with fast-growing poplar trees. You can harvest them and turn them into firewood logs, plus sawdust, twigs and leaves. You can also burn the sawdust, twigs and leaves, perhaps in a fluidized-bed combustor. Major pulp and paper companies like Ahlstrom, Kymmene and others, in Finland, Sweden, Scotland, New England and elsewhere now routinely use the wood residues from their plants to generate their own process steam and electricity. Alternatively, you can make methanol, known to generations of school-age chemists as 'methylated spirits'; its old popular name 'wood alcohol' correctly identifies both its classical origin and its chemical nature. It is an excellent liquid fuel, and has powered many Formula 1 cars around the world's Grand Prix circuits. If you prefer your biofuel gaseous, you put the pulverized wood into a gasifier akin to the ones mentioned earlier for coal. Biomass is easier to gasify than coal; and a biomass gasifier produces a fuel gas eminently suitable for burning, for instance, in a gas turbine or combined-cycle plant. Biomass gas-turbines now figure prominently in the future energy plans of Sweden; their potential worldwide is enormous.

As well as trees of many different types, grains like maize, sugar cane and many other agricultural crops are promising sources of bioenergy, not only the crops themselves but the residues from processing them – straw, husks, nutshells, waste cane or 'bagasse', the list goes on and on. Crops like these may be grown either for food or for energy, or indeed for both. For example, maize or sweetcorn has long been an important crop in North America, mainly as food for people and animals. It has also been popular in certain outlying areas as the feedstock for 'corn liquor', the untaxed and therefore illicit 'moonshine' of song and story. The active ingredient of 'moonshine' is ethanol or 'grain alcohol' – the alcohol of every 'alcoholic drink', produced by the age-old process of fermentation. Ethanol is another excellent liquid fuel, already being sold as 'gasohol' in the US Middle West, and vigorously supported as a substitute for vehicle petrol in Brazil; most of Brazil's cars now run on ethanol made from sugar cane. The current economic status of ethanol as a fuel is of course affected profoundly by world oil prices; when petrol is cheap ethanol is uncompetitive. The US gasohol programme has been criticized as just another farm subsidy. Brazil's indigenous ethanol programme has become hotly controversial. Nevertheless, the original purpose of the programme – reducing dramatically Brazil's vulnerability to the cost of imported petroleum – has been achieved.

Liquid fuels for transport are the most demanding of all categories of fuel; they must be portable and readily combustible but reasonably safe to carry, and they must concentrate a large amount of energy into a compact volume. Biofuels like methanol and ethanol contain somewhat less energy per litre than petrol; in other respects, however, they can do the job just as well as fossil fuels. In one major respect, indeed, they can do better. Ethanol, for instance, burns far more cleanly than petrol, producing no unburned hydrocarbons; and the carbon dioxide it releases is merely being returned to the atmosphere a year or two after being absorbed by green leaves.

Even when a crop is harvested purely for its food value, the processing always yields residual biomass, which may well outweigh the edible portion of the crop. Such crop residues represent a valuable source of bioenergy, either for direct combustion or for conversion to liquid or gaseous biofuels. Denmark, for instance, has banned farmers from burning straw in the fields; instead it is collected to fuel an expanding array of small local combined heat and power stations. Direct burning of crop residues has, however, one major drawback: it removes nutrients and organic material from the soil. Other options – in particular fermentation – may be preferable.

The burgeoning interest in growing plants explicitly for fuel raises, to be sure, several significant questions of policy. Should land that might grow food be turned over to fuel crops? In some areas of the world, where food is in short supply and good agricultural land at a premium, the question requires careful consideration. Biomass plantations need not, however, use good argicultural land. Some fast-growing trees and grasses suitable for bioenergy thrive on marginal land that would be unsuitable for food-crops. Moreover, as we shall see in the next chapter, more efficient use of bioenergy may actually enhance food production. Biomass fermentation in a 'digester' or 'biogas generator', for instance, not only produces usable fuel-gas but leaves behind a residue containing the nutrients and much of the organic matter, which can be returned to the land.

The 'digester' that extracts useful energy from biomass, and produces a residue still useful, need not of course be a steel or concrete tank. A cow or goat is a biomass harvester-cum-digester of considerable efficacy. The useful energy it extracts produces milk and meat; and its residue, in the form of dung, contains yet more usable bioenergy. A sun-dried cowpat is a biofuel in its own right, a compact and convenient combustible briquet highly valued in the rural Third World. But burning dried dung depletes the soil. As we shall see in the next chapter, using dung to produce biogas is both more efficient and more sustainable. In industrial countries, especially where livestock are kept in large numbers, the output of dung by the tonne can be a serious pollutant, overloading waterways and exuding an intrusive and unwelcome smell. An increasing number of farms have installed their own digester-generators, which produce usable biogas for the farm while converting the dung into an acceptable and innocuous soil-conditioner. While natural gas is cheap, biogas may not be economic. However, for the many farms unconnected to natural-gas mains, biogas from livestock dung could be a valuable energy resource in its own right.

Another biogas suddenly attracting enthusiastic attention comes from a source fully as unappetizing as livestock dung: rubbish tips. Modern industrial society generates a staggering mass of refuse; a single household can discard several tonnes a year. Much household refuse is in fact biomass, especially food waste and old newspapers and magazines; and this biomass has been accumulating for years in what the authorities euphemistically call 'landfill': in Britain more graphically known as refuse tips, in the US garbage dumps. Whether or not this is a sensible way to sacrifice usable resources – and many would insist that it is not – the industrial world's refuse tips already constitute a vast array of de facto biomass digesters, generating a steady exhalation of methane. Every now and then, the methane demonstrates its presence by exploding, sometimes with unfortunate consequences for neighbouring houses. Recently, however, enterprising groups have begun to drill 'gas wells' into well-aged landfills, extracting commercial quantities of methane from them – and of course reducing the possibility that the methane will burst out unbidden. The potential resource of landfill gas is startlingly large already, and easy to tap, although it also represents a potential hazard. The subterranean configuration of the average refuse tip is a lot less easy to map than gas-bearing geological strata, and the gas does not necessarily emerge only through the operator's valves.

WATER POWER

Of all the renewable energy technologies, water power has played much the most important role in modern industrial society. By any criterion, water power in its most conspicuous modern manifestation – large-scale hydroelectricity generation – is 'conventional'. As noted earlier in this chapter, however, modern hydroelectricity, although 'renewable', is by no means benign and free from undesirable environmental effects. Indeed a major hydroelectric source may not in practice be indefinitely 'renewable' either. A large dam slows down the river upstream, allowing solid silt to settle out onto the floor of the reservoir; siltation raises the floor of the reservoir, reducing the slope of the riverbed and slowing down the rate of flow into the reservoir. This in turn gradually diminishes the rate at which water can flow through the turbines, and hence diminishes the maximum output from the hydro station. The Aswan High Dam in Egypt, for instance, has for years been collecting behind it the silt that used to enrich the downstream banks of the Nile when the river overflowed each spring. Not only is the dam silting up, but the floodplains have lost much of their previously abundant fertility. Siltation is now also presenting a problem for the network of major dams run by the Tennessee Valley Authority in the eastern US.

Major dam projects have recently become hotly controversial, precisely because of their environmental impact. The Gabcikovo-Nagymaros project on the Danube, involving dams in Hungary and Czechoslovakia, was abandoned in mid-construction in 1989, when the Hungarian government yielded to domestic and international opposition and conceded that neither its economics nor its environmental side-effects could be acceptable. The Czech, Austrian and Hungarian governments are now embroiled in a fierce wrangle about what to do about the partially-finished structures, and who is to pay whom how much as a result. Turkey's Ataturk dam on the Euphrates, close to Turkey's border with Syria, has provoked bitter controversy with both Syria and Iraq, who claim that by damming the river Turkey is depriving their arid regions of water to which they are entitled. Indeed, Syria's own Tabaq dam now regularly finds that its turbines are out of the water. In India, the World Bank is caught up in a mounting furore about the grandiose Narmada dam project.

Nevertheless, even as mega-projects like these founder, less imposing hydro developments are once again finding favour – two approaches in particular. One is called, with simple accuracy, 'small hydro', the other 'run-of-the-river'. Each offers distinct advantages. Neither entails flooding major land areas; indeed, in appropriate locations, neither even requires a dam. Small hydro stations are sometimes subdivided into 'mini' and 'micro' hydro, depending on their maximum output; a mini-hydro station may have an output in megawatts, a micro-hydro station less than a megawatt. This output is to be sure tiny compared to that from a major scheme. However, locations suitable even geographically for major schemes are scarce – more so if political and ecological consequences are given due weight. A single massive installation takes a decade or more to construct, costs a staggering capital sum, inundates a vast area, and requires heavy-duty long-distance transmission lines extending many hundreds of kilometres. On the other hand, locations suitable for mini and micro-hydro installations abound. The small hydro alternative implies hundreds of modest installations, each blending easily into the terrain, diverting falling water from riverflow through pipelines to small turbines or bucket-wheels to generate electricity close to users. A unit can be built mainly in a factory, to a standard design – perhaps even delivered to the site by barge; and it can be delivering power and paying off its cost within a couple of years from the initial order. Once in place it can continue generating electricity for many decades, with only routine maintenance.

A site suitable for small hydro based on conventional water-turbines or bucket-wheels should have a significant gradient, allowing the water to fall far enough to deliver a worthwhile amount of energy for conversion to electricity. Another approach, suitable for more gradual gradients, is to use so-called 'bulb turbines'. Unlike a conventional water turbine, through which water falls vertically, a bulb turbine is mounted on a horizontal axis, with the generator enclosed in a sealed 'bulb' in line with the turbine. The river flows directly through the turbine, in one end and out the other, spinning the turbine as it flows through. A bulb turbine is therefore also called a 'low-head' turbine, because it operates without requiring a 'head' of water backed up behind a high dam to provide a long vertical drop. A bank of bulb turbines bridging a river can extract energy from the riverflow without making the river overflow its banks: a so-called 'run-of-the-river' power station. The dam serves only to support the turbine-generators, and direct the water-flow through them. France, for instance, has a large number of small run-of-the-river units, which make a valuable contribution, in flexibility and reliability, to the country's electricity supply.

A remarkable recent example of the potential of the concept is the

192-megawatt Sidney A. Murray power plant 320km north of New Orleans. The entire power plant, including eight bulb turbine generators, was constructed in a New Orleans shipyard in less than three years. When complete, it was floated up the Mississippi on a barge, and installed at the prepared site. Ordered in December 1986, it was operational in 1990.

The versatility of small hydro means that it could also make a significant contribution to electricity supply in many Third World areas.

WIND ENERGY

Sunlight energy transferred to moving water is comparatively easy to collect and convert, sunlight energy transferred to moving air somewhat less so. Moving air is less dense and massive than moving water, and therefore carries less energy in a given volume. Falling water will turn a turbine; falling air is perceptible only to weathermen and glider pilots. Moving air – wind – actually blows in three dimensions; but only the energy of its horizontal motion close to the ground is readily available for collection and conversion. A 'wind turbine' is therefore a first cousin of the run-of-the-river bulb turbine used for water-power. Wind turbines, however, must intercept a much larger total cross-section of the passing airstream to collect the same amount of energy. A wind turbine can collect at most about two-thirds of the energy in a stream of wind; the reason is obvious, if you reflect that collecting all the energy would bring the moving air to a standstill, leaving it stationary downstream of the turbine. Riverbanks keep water flowing along the same track; a water turbine does not have a swivel to present its broadest cross-section to the stream. Wind, however, can change direction continually; to maximize the energy it collects, a wind turbine may have to do likewise. Alternatively, it may be designed with a vertical axis, so that the turbine-blades catch the wind whatever direction it is blowing.

As you might expect, the faster the wind, the more energy it carries, over a disconcertingly wide range. The difference between a gentle zephyr and a hurricane is so enormous that no wind turbine could cope with both. Accordingly, wind-energy engineers have to collect detailed data on wind-behaviour – especially average and maximum velocities and variability – at a particular location, and design a wind turbine that will extract energy effectively, while withstanding the most severe blow that may strike it.

From the aftermath of the 1973 oil shock onwards, wind energy has excited growing enthusiasm in many parts of the world. Some early enthusiasm overreached itself, and led to manifestations like the 'Grosse Wind-Anlage' or Large Wind Plant – usually known as Growian – in Federal Germany. Growian never functioned properly. Eventually abandoned, it was a 'demonstration' of wind power that its supporters could have done without. Other demonstration units also encountered problems, with unsuitable blade designs and materials below specifications, requiring expensive maintenance and frequent shutdowns. Grandiose schemes for vast 'wind parks' sprang up and dissipated, damaging the credibility of the technology. The lessons, however, were learned. Generous tax breaks, notably in California, fostered the establishment of commercially-successful designs of wind turbine-generators. Although many of the early installations in the Altamont Pass in California did not live up to expectations, more recent units are generating electricity reliably and cheaply; the same is true of the spreading arrays of wind generators in Denmark.

Recent analyses indicate that in Britain, whose winds carry more concentrated energy than anywhere else in Europe, wind energy using proven designs could provide the cheapest electricity available. Britain's electricity supply industry remains, however, unconvinced. It has erected trial installations in Orkney, Kent and Carmarthenshire, and announced plans for three 'wind farms' – arrays of wind generators. It has also expressed interest in putting wind arrays offshore, perhaps in the North Sea. But it continues to warn about the difficulties of 'public acceptance' of wind arrays, arising from the visual intrusion and the noise they create. Wind-energy advocates insist that the noise is not much more pronounced that the noise of the wind itself. They point to the visual intrusion of massive nuclear and coal-fired power stations, and the 49 000 pylons already strung out across the British countryside. They suspect that the concern about these issues in the context of wind energy is prompted primarily by the industry's desire to keep wind energy from intruding on traditional plans and policies.

Be that as it may, wind energy is now technically feasible and economically competitive in at least some areas of the world, depending particularly on the current local status of natural gas for electricity generation. Wind generators can be readily fabricated on an assembly-line basis, transported in 'kit' form and erected on prepared sites. As construction facilities become more widely established, unit costs will certainly come down further. Wind energy is also inherently dispersed, not centralized. In rural areas, including many parts of the Third World, wind resources may therefore prove to be a useful supply technology even far from electricity grids.

WAVE ENERGY

No matter how subtle and sophisticated the designs of wind turbines become, nor how much they proliferate, their capacity will never equal that of the world's largest wind-energy collector: the surface of the ocean. Wind rushing along an uninterrupted stretch of open ocean transfers energy into the uppermost layer of water. As the wind drags each bit of water in its path, the water finds the next bit of water in the way; the result is a rolling wave, advancing along the water, sometimes for many hundreds of kilometres, gathering energy as it rolls. Think of wind-ripples on a pond, and multiply the effect many millionfold.

Eventually the rolling ocean wave reaches shallow water near land. By this time, each metre of advancing wavefront may be carrying more than a million joules per second – that is, a power of more than a megawatt. Friction with the seafloor bleeds this energy away, until the last of it dissipates in breakers crashing against seacliffs, or surf surging up a beach. The energy content of sea-waves can be awesome, and accordingly tempting; but its very abundance makes wave-energy a daunting challenge to harness.

Even so, the oil shock of 1973 spurred a burst of enthusiasm for wave energy, notably in Britain. A report from the Central Policy Review Staff to the British Cabinet identified wave energy as the most promising of all the renewable energy options. Britain was already involved in offshore engineering for North Sea oil and gas. With considerable fanfare the government launched a programme of research and development into suitable technologies for wave energy. For several years possible wave energy concepts proliferated, of which probably the best known was the so-called 'duck' conceived by Stephen Salter of the University of Edinburgh. Tests on prototypes gave encouraging results; and careful costings suggested that wave energy could generate electricity at least competitive with that from nuclear power. In the 1980s, however, the government's Advisory Committee on Research and Development for Fuel and Power, for reasons that have never been adequately explained, recommended that the programme be abandoned. Wave energy people remain convinced that the decision was taken to defend Britain's struggling nuclear power programme from competition.

Norway, too, initiated a programme of wave energy research; a prototype generator on the Norwegian coast was destroyed by a violent storm in the northern North Sea, but work is continuing. Japan, too, has maintained a modest programme. A design team based at Queen's University, Belfast is installing a small unit on the island of Islay off the west coast of Scotland. Plans for wave energy generators are also under consideration for sites in Indonesia. The resource is certainly there. If engineers can devise a sufficiently robust system to collect it, wave energy could yet make a useful contribution to electricity supplies in many areas of the world.

TIDAL ENERGY

Waves just advance. Tides, however, advance and retreat; and the moving water carries energy, no matter whether it is coming in or going out. Wave energy comes from the wind, and thus originally from the sun. Tidal energy does not. The gravitational pull of the moon makes the fluid oceans bulge toward it. The solid earth rotates under the bulge, buffeting it back and forth between continents, making water levels on some coastline rise and fall with rhythmic predictability, according to the shape of the coastline, the position of the moon and the local time of day. The oceans sloshing back and forth on the surface of the earth create a turbulent frictional drag that is gradually slowing down the earth's rotation – very gradually, to be sure. The energy of tidal flows is a manifestation of this relative motion between the solid earth and its watery coating.

Tidal energy is worth collecting only if the water surging back and forth about twice a day carries enough energy to turn turbines – usually bulb turbines, like those in a run-of-the-river hydro station. The greater the local rise and fall – the 'tidal range' – the more tidal energy available. But the ebb and flow must be channelled through the turbines. An appropriate site for a tidal power station is therefore a narrow bay or inlet with headlands to either side; the dam to carry the turbines can then be built from one headland to the other, across the tidal flow. Two engineering problems arise. One is that half the time the tidal current is flowing toward the land, half the time away. The other is that the tidal current varies steadily, from essentially zero at low tide, when the tide is turning, to a maximum three hours later, and back to zero at high tide, before reversing direction. Moreover, the tidal range itself varies significantly through the year. When the moon is full or new, and in line with the sun, their gravitational attractions reinforce each other, making the oceans bulge more and producing extreme 'spring tides'. At half-moon, however, with the moon at a right angle to the sun, their gravitational attractions offset each other, producing shallow 'neap tides'. Although the output from a tidal power station can be predicted almost as precisely as the tides, it will vary a great deal, not only month by month but hour by hour; indeed every six hours it will drop to zero, unless engineers arrange to store and channel water to smooth out the tidal variations.

Accordingly, tidal power stations require elaborate civil engineering, to steer waterflow in the desired direction, and to retain water in one or more reservoirs or 'basins', so that turbines keep turning even when the tide itself is turning. The world's only operating tidal power station of significant size is at La Rance on the Brittany coast of France. It was built between 1961 and 1967, as a 240-megawatt prototype for a series of larger stations – which never, however, materialized, as the French nuclear power programme expanded. The 24 10-megawatt bulb turbines at La Rance can generate electricity on both ebb and flow; they also act as pumps, driven by electricity from the French grid, to shift water from one side of the dam or 'barrage' to the other, to store it for release again at times of high electricity demand.

In Britain, the tidal surge up the Severn River has tempted engineers since as far back as 1849. Proposals for a Severn Barrage continue to excite both enthusiasm and controversy. Proponents have presented detailed designs and carried out extensive feasibility studies, indicating that such a barrage could supply some 10 per cent of the electricity used in Britain. But the initial capital cost would be well into the billions of pounds, and the timescale well over a decade. Moreover the impact of a barrage on the ecology of the Severn Estuary would be unpredictable, and could be seriously damaging.

The Bay of Fundy in Nova Scotia has the largest tidal range in the world – as high as a startling 17 metres. It, too, has attracted engineers for many decades. Again, however, the capital outlay would be daunting, as would the actual engineering, carrying out construction work amid so much untrammelled water-power.

The basic engineering for tidal power is much the same as that for hydro power. But the geographical prerequisites for a tidal power site, and the inevitable variation in the energy available for conversion, mean that tidal power will always be at best a minor and local contributor to the global energy mix.

GEOTHERMAL ENERGY

Another possible energy resource unrelated to sunlight is the heat within the earth – so-called 'geothermal' energy, which is released by radioactive materials far below the earth's surface. Its most spectacular manifestation is a volcanic eruption. No one – at least no one rational – has suggested attempting to harness a volcano; but geothermal energy also emerges at a more manageable rate in many places where the earth's crust is thin enough. Iceland, for instance, has more than its share of active volcanic sites; so does the central North Island of New Zealand. In both areas, boreholes drilled to moderate depths tap into subterranean reservoirs not of water but of steam; the steam is piped directly into homes in Reykjavik and Rotorua to provide heating and hot water. New Zealand also has one of the few electricity-generating stations powered by geothermal steam, at Wairakei. Other well-known geothermal power stations are at Lardarello in Italy and the Geysers in California.

Only a limited number of accessible reservoirs of subterranean steam or hot water have been located to date. But the subterranean rocks themselves may be hot enough to heat or indeed to boil water injected down a well from the surface. So-called 'hot dry rock' geothermal energy is now believed to be available in many parts of the earth. In the suburbs of Paris, for instance, blocks of flats are now heated by geothermal energy. In Britain, plans were well-advanced for a hot dry rock geothermal installation at Southampton in the early 1980s, until the Central Electricity Generating Board abruptly withdrew its support.

Geothermal energy is not, in practice, strictly 'renewable'. As a borehole extracts heat from a subterranean reservoir the reservoir gradually cools; heat does not flow into it from its surroundings rapidly enough to replace the heat extracted for use. Moreover, the steam or hot water from classical geothermal sources like Wairakei or the Geysers brings up from underground also noxious gases like hydrogen sulphide that give the neighbourhood the unmistakable whiff of rotten eggs, and may be serious local pollutants. The mineral content of the discharge can affect not only pipework and other hardware but also local waterways. For hot dry rock geothermal energy, however, these problems are less significant, if they arise at all; and its potential abundance, for low-temperature heat for district heating if not for electricity generation, ensures that energy planners will do well to keep it in mind.

SOLAR ENERGY

Jonathan Swift, ridiculing science and scientists, had his hero Lemuel Gulliver visit the wise men of the airborne floating island of Laputa. One 'had been eight years upon a project for extracting sunbeams out of cucumbers, which were to be put into vials hermetically sealed and let out to warm the air in raw inclement summers'. We can do better. Biomass fuels, hydroelectricity and wind power are all solar energy in disguise. But what about the real thing? Instead of collecting solar energy only at second hand, from green leaves, falling water and moving air, what about collecting the sunlight itself? To be sure, as already discussed in Chapter 2, solar energy already makes the world go round. Moreover, as we saw in Chapter 7, effective building design – so-called 'passive solar' design – capitalizes on the available solar energy for both light and warmth. Can we, however, capture solar energy explicitly and directly, for active intentional use? Provided we do not want bottled sunbeams, but will settle for them converted into heat or electricity, the answer of course is yes.

Whenever the sunlight is bright enough to read by, solar energy is available for collection, no matter whether it feels warm. What you are collecting is not the perceptible heat but the 'brightness' – the visible radiant energy. The solar energy-flow on a cloudless winter day in Scandinavia, for example, should not be undervalued; it may be fully half of that on a summer day in Arizona. How you collect solar energy depends on what you want to do with it. As usual, the easiest task is to raise local temperature a few tens of degrees, for space heating or hot water. A 'solar collector', essentially a flat plate of absorbent black metal or plastic, carrying watertubes and covered by clear glass or plastic, will absorb enough bright sunlight to heat water nearly to boiling point. The hot water can be piped by its own convection currents through radiators, or stored in a tank to be tapped for baths and washing. Travellers in Mediterranean countries like Greece and Israel now see the inclined planes of solar collectors on roof after roof, often with a water-cylinder attached, even in the most remote rural areas. In California and other sunny states, swimming pools are now routinely plumbed into solar collectors to keep them comfortable the year round.

Some houses and commercial buildings in colder climates have incorporated much larger water tanks, often underground and thickly insulated, able to store enough solar heat through the summer to keep a thermally tight building comfortable for an entire winter. For example, the headquarters building of Ontario Hydro, a massive office block in Toronto constructed in the 1970s, includes a water tank that can store not only solar heat but also that given off by occupants and office equipment. Automatic controls circulate the water to stabilize indoor temperatures whatever the weather outside – a striking innovation from a major electricity supplier. At present, however, especially when natural gas prices are low, so-called 'active' solar space heating, involving dedicated plumbing and water-circulation, is less attractive than 'passive' solar heating that relies on the building architecture itself. Solar space heating, whether passive or active, is also of limited value for existing buildings; improved insulation is much more important and effective. In the longer term, however, buildings with integral water-circulating systems will be able to gather and use heat from any source available, including not only solar collectors but also, for instance, local district heating grids. Kungalv, a town in Sweden with 12 000 inhabitants, has district heating provided by 100 000 square metres of solar collectors on a nearby hillside, coupled to a rock cavern to store the hot water for use in winter.

Industry uses low-temperature heat not only for space-heating but also in industrial processes. Even a simple solar collector can warm feed-water sufficiently to cut fuel bills significantly. More sophisticated collectors, using focusing reflectors to concentrate the sunlight, can produce high-quality steam. At a solar energy research station built at Odeillo in the Pyrenees in the late 1960s, the entire side of a tall building was shaped into an enormous curved reflector; the solar radiation concentrated at its focus was intense enough to melt steel armour-plate. In the Mojave desert near Barstow in California, the 'Solar One' power plant came into operation in 1982. An array of 1800 computer-controlled mirrors or 'heliostats' track the sun and direct its rays onto a bulb-shaped tank atop a central 'power tower', producing steam good enough to spin a turbine and generate 10 megawatts of electricity. In 1989 Luz Engineering received an order from San Diego Electric for an 80-megawatt power plant of this so-called 'solar thermal' design. Elsewhere, however, the solar thermal concept has encountered engineering difficulties, and its cost remains too high. Thus far it has failed to make headway against low-priced electricity from conventional technology. The Themis demonstration project in southwestern France, launched in 1975, was originally intended to produce 10 megawatts of electricity for Electricité de France; but its output had to be reduced to only 2.5 megawatts, it suffered wind damage and eventually faded out.

In general, nevertheless, gathering sunlight energy to use as heat is both straightforward and technically easy. In purely physical terms, however, it is also wasteful, since sunlight energy is inherently of much higher quality than heat energy. The real prize is to gather sunlight energy in a form that preserves its inherent high quality: to gather it, for instance, directly as electricity. Certain natural materials do this already; the retina of your eye, for instance, responds to the sunlight falling on it by generating a minute electrical current that your brain interprets as 'seeing'. In an 'electric eye' – more scientifically known as a 'photo-electric cell' – light falling on a material like the element selenium dislodges electrons from its atoms, producing a small electric current.

The materials that can convert sunlight directly into electricity on a significantly useful scale are artificial 'semi-conductors', with distinctive electronic properties akin to those of transistors and microchips: in particular silicon crystals, 'amorphous' silicon, and gallium arsenide. Whichever material is chosen, the detailed structure of the resulting 'solar cell' is simple but subtle. A thin layer of the material is sandwiched between an upper conducting 'grid' – openwork so as not to obstruct the arriving sunlight – and a baseplate that provides the other end of the electric circuit. Sunlight penetrating the material dislodges electrons, which drift across a junction-layer in the sandwich and set up an electrical pressure or 'voltage' between the grid on top and the lower baseplate. This voltage makes an electric current flow through a wire or other conductor connecting grid and baseplate: the more sunlight, the more electric current. The concept is called 'solar photovoltaics', a mouthful often abbreviated just to PV.

Many analysts now believe that biomass and photovoltaics, supplying fuels and electricity respectively, will be the most important innovative energy supply technologies in the coming century. In the meantime, however, PV still has major hurdles to overcome. As already noted, sunlight is a diffuse and variable source of energy. A solar cell, even of the best design available today, can convert at best less than 30 per cent of the incoming sunlight energy into electricity; and designs that are cheaper to fabricate are much less efficient. The cost of electricity from an array of solar cells is usually given in 'dollars per peak watt' – that is, the cost of the array in dollars divided by the maximum electrical output in watts that it can produce in the brightest available sunshine. Solar cells now commercially marketed cost of the order of five dollars per peak watt. Comparison with a traditional electricity supply technology like a coal-fired power station is difficult; it involves a number of assumptions about the operating lifetime of the array and of the power station, the variation of sunlight, the future cost of coal, interest charges on capital and other relevant details. But the conclusion generally reached is that the cost of PV must come down to about one dollar per peak watt before PV can be considered economically fully competitive.

The cost of PV has, however, been dropping dramatically. In the 1960s and 1970s solar cells were made from thin slices of silicon crystal, with precise human craftsmanship. An array large enough to supply a useful output of electricity was so expensive that it could be used only in exotic contexts, like the Apollo moon missions and the Skylab satellite. Reducing the cost entailed automating the fabrication process; and that in turn meant designing a cell that could be manufactured in an automated plant. A major breakthrough came with the discovery that silicon with suitable properties did not have to be sliced meticulously off a large single crystal, but could be deposited as a thin 'amorphous' – non-crystalline – layer on a backing like glass. In the US, for instance, Solarex, one of the major manufacturers of solar arrays, is constructing a fully automated computer-integrated manufacturing plant able to turn out about one megawatt per year of amorphous silicon PV arrays. A further plant with an output of 10 megawatts per year of more powerful PV arrays should bring the cost down to about one dollar per peak watt.

Critics of solar energy often point to the need to allow diffuse sunlight to fall on a large area to collect enough energy to be of interest. Some proposals suggest, for example, siting vast arrays in deserts like the Mojave in California or the Sahara in Africa, to produce outputs like those of conventional baseload power stations. Another alternative, however, may make better sense, at least in the short term. In any sizeable city, sunlight falls on many square kilometres of roof. Every building has a roof; and almost the whole of the roof-area is both unoccupied and unused. As the cost of PV comes down, why not make at least part of the roof out of PV arrays?

New England Electric is already investigating the performance and potential of PV roofs. Its Gardner PV Project has installed PV roofs on 30 homes; each roof is wired through the household electricity metre into the New England Electric grid. At night, or when the sun is not shining, the household draws its electricity from the grid, and the metre turns to measure how much the family will have to pay. When the sun is shining on the PV roof, however, the household draws its electricity from its own roof; and any surplus is fed through the metre back into the grid – turning the metre backwards and reducing the family electricity bill. The Gardner project, although not at present economic, is intended to give the utility experience of integrating the concept into its system, to be ready when PV roofs can be installed at acceptable cost. Linking many households into the grid in this way eliminates what would otherwise be a key problem for PV: storing electricity from an individual PV array, to keep the lights on when the sun is not shining.

For an isolated PV array, the only available way to store solar energy to use at night or under a heavy overcast is a bank of batteries of one kind or another. Batteries themselves are expensive. To store enough energy for a prolonged dark spell means both a bulky and cumbersome battery bank and a substantial additional capital outlay. In the longer term, however, as cells become cheaper, PV electricity could be used to generate hydrogen, by 'electrolysis' of water, familiar from countless classroom experiments. The hydrogen could be stored, transmitted and used much like natural gas. In 1990 the World Resources Institute published a report by Joan Ogden and Robert Williams of Princeton entitled 'Solar Hydrogen: Moving Beyond Fossil Fuels', which explores the ramifications of this wide-ranging option in lucid detail.

The solar hydrogen option has had its advocates since the 1970s; but its feasibility now looks much more imminent with the falling cost of PV, and its potential importance much more profound as the environmental impacts of fuel use grow more onerous. Nevertheless, another school of thought would ask: why try to centralize an inherently decentralized energy source like sunlight? As we shall see in the next chapter, opportunities for using PV in small local applications abound, especially in the Third World. But experience suggests that – like any other energy technology – PV must be installed with a genuine understanding of how it is to be used and by whom, in what circumstances. Otherwise it may be just as misguided, extravagant and ultimately disruptive as any traditional energy technology.

THE ENERGY SHOPPING LIST

The shopping list for fuel and electricity supply technologies is already long, and growing longer. We can pick and choose from a wide variety of innovative options already known to be technically feasible, with environmental impacts that may be significantly more acceptable than those of traditional fuel and electricity supplies. But traditional fuel and electricity suppliers respond that many of the alternative supply options described in this chapter are 'not economic', and therefore uninteresting. Before we accept such dismissive economic comparisons, we must be sure of the ground-rules on which they are based.

As we have seen, traditional fuel and electricity supply technologies may bring with them serious and deleterious impacts on the environment. How do the economic comparisons take account of environmental impacts like acid rain, toxic wastes, and the greenhouse effect? Traditional fuel and electricity suppliers often receive generous financial subsidies, direct or indirect, from governments – that is, taxpayers – although ascertaining the details of these subsidies is frequently difficult if not impossible. Governments may offer tax breaks and allowances for investment in new supply facilities. Governments may guarantee loans to fuel and electricity suppliers from private banks and finance houses, so that the suppliers get access to capital on advantageous terms. Governments may offer grants for research and development on traditional fuel and electricity technologies; for example, even today studies of the environmental impacts of fossil fuels and nuclear power are financed substantially by government funds. Governments provide insurance against the consequences of nuclear accidents; and so on. Unless and until we can compare 'like with like', unless and until we can compare, say, fast breeder reactors and solar photovoltaics on equal terms in every respect – capital costs, running costs, environmental impact, and social and political implications – dismissing innovative options as 'uneconomic' is simply special pleading on behalf of an increasingly precarious status quo.

Innovative fuel and electricity supply options are also frequently discounted as unable to meet the kind of 'demand' projected by traditional forecasters, in which global use of fuels and electricity is expected to double or triple in the next three or four decades. This, too, is an unacceptably one-sided view: because using traditional fuels and electricity on such a scale would raise equally insuperable problems of capital cost, environmental impact and social and political implications. If, however, we redirect our attention first to improving the efficiency of the end-use hardware we use – buildings, lighting, heating and ventilation, industrial plant, vehicles and all the rest – we can reduce dramatically the amount of fuels and electricity we shall need to run them. At the same time we can gather the necessary experience of the innovative supply options, reducing their cost and improving their performance. Efficient end-use and innovative supply technologies are complementary. If we make the right choices, we can progressively shift the balance of our energy systems. We can provide energy services more reliably, and at lower cost, from fuel and electricity supply technologies whose side-effects are more tolerable, both environmentally and socially. To do so, however, will require concerted, continual effort – political effort, in the best sense of the term. The forces of tradition will not lightly surrender their prerogatives and their privileges.

9
POWER TO THE PEOPLE

Of all the energy systems thus far invented, which do you suppose is the most essential to the most people? Lightbulbs, motors, gas fires, cars, power stations – the shopping list gets ever longer; but the correct answer goes all the way back to when human beings first began using energy. Despite the dramatic advances described in previous chapters, the energy system most essential to most people on earth is still the bonfire. More people get their essential energy from bonfires than from all the world's coal mines, oil wells, refineries and power stations combined. That alone is a measure of how far we still have to go.

More than five billion people now live on our planet; by the year 2000 we may number six billion. The earlier chapters of this book have focused, however, almost entirely on circumstances in so-called 'industrial countries'. An industrial country is, among other attributes, a country in which people use energy to provide a vast range of services, more or less as a matter of routine. In general they have ready access to comfort, light, cooked food, mobility, and products that are processed, manufactured and transported using energy hardware, fuels and electricity – everything from skyscrapers to microchips. Even in industrial countries, to be sure, not everyone is so fortunate; and fewer than two billion people live in industrial countries. The other three-plus billion live in countries variously described as 'less developed', 'developing', or 'Third World'.

None of these labels is satisfactory; and applying a single label to countries as disparate as, say, Brazil and Bangladesh, or Malaysia and Mali, blurs distinctions of geography, resources and culture that may be profoundly important. Nevertheless, in what follows we shall refer – for lack of any more satisfactory designation, and somewhat apologetically – to the 'Third World', and hope the context makes the meaning clear enough.

The question of 'development' – what it means and how to achieve it – has already filled many bookshelves; it will fill many more. This chapter can touch on only one major aspect of 'development', albeit a crucial one: providing adequate energy services to the majority of people on earth who do not now have them – those in the Third World, and especially the rural Third World. As already indicated, Third World countries differ widely in geography, resources and cultures; no 'typical' Third World country actually exists. But some problems, including energy problems, are common to many Third World countries – particularly in the rural areas where most of the people still live, and where the bonfire is still the key to using energy.

URBAN SERVICES ENERGY

To be sure, providing energy services in Third World cities also raises problems – much like those in industrial cities, albeit often more acute. Air pollution, congestion and noise from cars and other vehicles powered by internal combustion engines are as punishing in Mexico City as in Manhattan; coal-burning urban factories and power plants in China blight the neighbourhood just like those in Cleveland. Since the 1950s, many Third World cities have come increasingly to resemble industrial cities. The reason is simple, and its implications dismaying: from the 1950s onwards, Third World planners took the existing industrial world as their model. They construed 'development' to mean sprinting as rapidly as possible along the path previously traced by the countries of Europe and North America, from the nineteenth century onward.

In Europe and North America, as agriculture became ever more mechanized, people moved from the country to towns and cities; manufacture evolved from craftwork to factories; and transport links for food and other commodities grew ever longer and more elaborate. All these developments entailed using more energy hardware, more fuels and more electricity. Many political leaders of Third World countries were educated in industrial countries, sometimes in the aftermath of colonial rule by Britain, France or another of the old 'imperial' industrial powers. Third World leaders sought to emulate the example of the existing industrial world, not least in providing energy services.

In Third World cities they constructed commercial and residential buildings almost identical to those in industrial countries, often regardless of local climatic conditions – in particular the geographical fact that Third World cities are almost invariably closer to the equator than most industrial cities. They reorganized urban communities to cater for motor vehicles. They established industries using processes originating in industrial countries. These policies in turn entailed providing fuels and electricity to run the buildings, vehicles and industrial plant. All the energy systems involved were modelled on examples in industrial countries, and indeed frequently manufactured by the same industrial suppliers. Third World planners also adopted traditional analytical methods for studying and forecasting future energy use in their countries.

Unfortunately, however, the period of rapid 'modernization' in many Third World cities coincided almost precisely with the period of very low world oil prices, between 1950 and 1973. As a result, Third World cities acquired the sort of energy hardware – buildings, vehicles, the industrial plant – that industrial countries were using when its costs and the environmental impacts of fuels and electricity went almost unnoticed. Such hardware is likely to be inefficient and polluting, a major liability as fuel costs rise and environmental standards stiffen. A glass-and-concrete office block built in Frankfurt or San Francisco in the 1960s requires heavy-duty heating and air-conditioning to compensate for the shortcomings of the structure; a similar office block in the climate of Nairobi or Buenos Aires may demand yet more fuel and electricity even to make it habitable. One skyscraper in Caracas, Venezuela, built on the industrial-world model with glass curtain walls, is exposed to such fierce sunlight that the windowpanes of the upper stories expand in the heat, shatter and rain shards onto the pavements far below.

In their rush to develop, Third World cities all too often mimicked the most extravagantly wasteful excesses of the industrial world. Governments and companies from the industrial world actively supported and encouraged these policies, not through any lack of scruple but merely because such an approach was accepted and advocated almost universally. In Third World cities as in industrial cities, planners construed the demand for comfort, nourishment, illumination, mobility and other essential energy services as a demand for fuels and electricity, to run whatever energy hardware happened to be installed. Until 1973 the approach was as understandable and defensible in New Delhi as it was in New York. Thereafter, however, it brought in its train problems that proved even more acute in Third World cities than in industrial cities.

In 1973, as we have seen, OPEC shook the industrial world to its foundations by quadrupling the world price of oil. For those few Third World countries fortunate enough to have their own oil reserves, the quadrupled price brought a windfall of income that financed a rush of capital investment in places like Kuwait and Saudi Arabia. For Third World countries without their own oil, however, the price shock of 1973, and the further shock of 1979, were brutal and crippling. Suddenly they found that their export earnings, largely from basic bulk commodities like sugar, tea and coffee, were being devoured by the bills they had to pay for high-priced imported oil to run their inefficient buildings, cars, factory boilers and power-plants. Many Third World countries plummeted into ever deeper debt; and capital investment of any kind, including that for energy services, became acutely difficult. In the 1970s and 1980s a rising tide of people poured out of rural areas into the cities in search of a better life. What they found all too often was that they had only exchanged rural poverty for urban poverty, squatting in 'barrios', 'favelas' and other miserable shantytowns that sprang up in many Third World cities. The surge of urban populations imposed a further burden on services, including energy services, already struggling to meet even basic needs.

Because of their comparative proximity to the equator, most Third World cities have less need for low-temperature heat to make buildings comfortable. On the contrary, the need is for air-conditioning – that is, cooling – especially in buildings constructed like those in the industrial world, whose structure does at best a mediocre job of isolating the indoors thermally from the outdoors. Almost all air-conditioning hardware commercially available needs electricity to run it. In the Third World, electricity supply planners expect their peak demand on scorching summer afternoons; the result all too often is power cuts, 'brownouts' or complete failure of electricity supply just when it is most needed.

Electricity grids in Third World countries have broadly followed the pattern of development demonstrated in the industrial world, beginning with small local stations, often powered by diesel engines, and moving on to larger traditional coal-fired and oil-fired stations with boilers and steam turbines, interlinked with high-voltage transmission lines. Many Third World countries also have significant hydroelectric capacity, often supplied by enormous dams built with financial support from the World Bank and other development funding agencies. In the early 1970s, a number of Third World countries also imitated the industrial world by announcing vast plans for nuclear power programmes.

In 1975, for example, in the largest export contract ever signed, Federal Germany agreed to supply to Brazil not only eight large pressurized-water nuclear power stations but also a uranium enrichment plant and a spent-fuel reprocessing plant. In the event, only two nuclear power stations even started construction; neither is yet complete fifteen years later, and many observers expect the second to be abandoned unfinished. The Shah of Iran had even more ambitious plans, for twenty nuclear power stations in operation by the year 2000. His fall, however, also terminated work on the only two units under construction; the half-finished hulks at Būshehr remain as a monument to grandiose aspirations that went astray.

Nuclear programmes in the newly industrializing countries, notably South Korea, fared better. In general, however, the high hopes of the reactor manufacturers for a burgeoning export market in the Third World have withered. Several Third World nuclear establishments, like those of India and Pakistan, nevertheless retain an influence far out of proportion to their actual achievements. The performance of India's Department of Atomic Energy in building its series of nuclear power stations has been lamentable; but it continues to map out vast future schemes, while siphoning off a wildly disproportionate share of India's limited resources of capital and skills.

Pakistan, for its part, has a single small and elderly nuclear plant of the Canadian CANDU design, near Karachi; but it too insists on the importance of nuclear power for electricity supply, and has built a top-secret uranium enrichment plant at Kahuta. The Karachi reactor, oddly enough, does not require enriched uranium fuel. Neither Pakistan nor India is a party to the Non-Proliferation Treaty, according to which more than 140 countries of the world have agreed not to acquire nuclear weapons. Despite the protestations of the Indian and Pakistani governments, outside observers continue to believe that both these countries, and some others, look upon their nuclear programmes as at best only partly civil in aim.

As regards functioning electricity supply in return for expenditure, however, the nuclear programmes have been not an asset but a liability. The World Bank has never financed a nuclear power project, for purely economic reasons. Throughout the 1970s and 1980s, on the other hand, the World Bank and other international funding agencies supported a variety of other electricity supply projects in Africa, Asia and Latin America, including hydroelectric dams, diesel and coal-fired power stations, gas turbines and transmission lines. In the 1990s, Third World electricity-supply projects are expanding more rapidly than ever, especially in southeast Asia. Yet no matter how rapidly Third World electricity supply capacity expands, electricity use, especially in cities, is trying to expand even faster. Would-be electricity users switch on more and more appliances, overloading generating plants and transmission and distribution systems, until circuit-breakers plunge the area into darkness and the air-conditioning stops.

Attempting to keep up with the accelerating demand for energy services powered by electricity, Third World planners are caught up in a frenzied rush to build new generating capacity and improve the performance of existing plant. India's National Thermal Power Corporation, for instance, has already embarked on plans for an awe-inspiring series of more than 30 large coal-fired power stations to be commissioned by 1995; and India's Power Finance Corporation is expecting World Bank support for refurbishing and modernizing a further 34 thermal stations, including a total of 162 units, as well as 49 small hydro stations. Electricity suppliers in Thailand, the Philippines, Pakistan, Indonesia and elsewhere in southeast Asia are confronting similarly daunting increases in demand for electrical services. Elsewhere in the Third World, the demand is not growing quite so fast; but it is still outstripping existing and projected supplies. Technical problems often complicate matters further. In Argentina, for example, the country's three nuclear power stations are prone to unplanned shutdowns; and the huge hydroelectric station at El Chichon has been severely crippled by the discovery of a crack in its towering dam.

The implications of the soaring Third World demand for electricity reach their most alarming, however, in the People's Republic of China. At the end of the 1980s China became the world's largest producer of coal, extracting more than 900 million tonnes a year – almost one-fifth of global coal production. A brief spasm of expectation that China would enter the world coal market as a major exporter rapidly faded. Instead, almost the whole of China's vast output of coal is being burned in China's own power stations and industrial boilers, most of which are frighteningly inefficient and devoid of any form of pollution control. The capital cost of upgrading literally thousands of coal-fired plants to meet even minimum standards of efficiency and environmental acceptability would be awesome. Yet China has embarked on a policy of providing its more than one billion citizens with amenities like electric refrigerators, which will require a further major increase in electricity supply within a matter of years – especially since the refrigerator design is itself inefficient. If the increase in supply comes from traditional Chinese coal-fired generating plants, the local impact of the sulphur and nitrogen oxides and the global impact of the carbon dioxide released may be devastating, whatever happens elsewhere in the world.

China has expressed enthusiasm for nuclear power, and is building nuclear stations at Daya Bay northeast of Hong Kong and at Qinshan near Shanghai; but the sheer capital cost of nuclear plant may be insupportable. China appears to have shelved the Three Gorges hydroelectric project on the Yangtse River, which at a proposed output of 15 000 megawatts would have been one of the largest in the world; but many smaller hydroelectric projects figure prominently in Chinese electricity plans. These plans call for a total installed capacity of 224 000 megawatts in operation by the end of the 1990s, more than doubling the 110 000 megawatts of capacity at the end of the 1980s. Analysts outside China see no way for China to achieve such a target in a decade.

The Chinese situation may appear an extreme example, but it is echoed in many other parts of the Third World, especially in countries where most of the people live in remote rural settlements, for instance African villages. In the traditional approach to electricity supply, modelled on that in industrial countries, rural populations come at the back of the queue, partly because they are spread comparatively thinly over a vast area, and partly because the problem of supplying densely-populated urban areas is already overwhelming. Electricity planners, abetted by the international funding agencies and plant suppliers from the industrial world, tend to focus their attention on major generating projects. Such projects can be centrally planned, financed and administered, and their substantial output can be delivered to cities, where the factories are, and where the well-off electricity users with the electrical appliances mainly live. Supplying electricity thus widens the already yawning gap between the small minority of 'haves' and the great majority of 'have-nots' in the Third World, aggravating tensions and exacerbating inequities.

Against such a background, traditional commentators from industrial countries dismiss out of hand the alternative strategy outlined in earlier chapters. They see, for instance, no possible role for improved efficiency in Third World cities: 'How can they conserve what they don't have?' Such criticism betrays yet again the fundamental misconception that 'energy' means 'fuel and electricity supplies'. Yet a key reason for the perpetual overload on Third World electricity supply systems is the low operating efficiency of air-conditioners, refrigerators, lights and other appliances using the electricity. The World Resources Institute study 'Energy for a Sustainable World' points out with blunt emphasis that the affluent elite in Third World countries, who account for most commercial energy use, waste at least as much fuel and electricity as energy-users in industrial countries. Their appliances are often less efficient even than those now on sale in the industrial world, partly because manufacturers frequently export old, inefficient stock to the Third World. Every time a well-to-do Brazilian, say, switches on a new refrigerator half as efficient as it might be, it uses twice as much electricity as necessary to provide the same service. To provide this extra electricity Brazil's electricity suppliers have to install twice as much capacity as necessary to generate, transmit and deliver it to the new refrigerator. When an electricity supply system is already overloaded, inefficient appliances impose an especially unwelcome burden.

'Energy for a Sustainable World' argues that – far from being irrelevant in the Third World – efficiency is even more important. Where resources, especially financial resources, are scarce, 'least-cost planning', aiming for maximum benefit from every dollar, rupee or cruzeiro spent on energy services, is crucial. A recent report from Lawrence Berkeley Laboratory in the US, entitled 'The Conservation Potential of Compact Fluorescent Lamps in India and Brazil', gives a striking example of what this approach could achieve. The report concludes that simply by shifting subsidies from electricity supply to end-use India could be saving US$1.2 million and Brazil US$2.5 million per day within a decade. India and Brazil already subsidize the supply of electricity to the poor, nowhere more than in lighting. Lighting sharply pushes up the peak load on the electricity supply system. In Brazil the residential daily load triples after 5 pm. In India, at 6 pm some 30 per cent of total incandescent lighting load is switched on in a quarter of an hour. Traditional planning deals with this phenomenon by building new electricity supply capacity, and subsidizing the cost of the electricity to users – in effect helping the supplier to finance the new capacity.

The Lawrence Berkeley study proposes that planners shift the subsidy, to finance the distribution of high-efficiency compact fluorescent lamps (CFLs) to users. According to the British newsletter 'Energy Economist', October 1989, such a shift in policy 'would yield truly staggering returns. Over ten years, if, as one scenario suggests, India had 20 per cent of lighting in CFLs and Brazil 36 per cent, then power utilities would be able to avoid building 8 gigawatts of capacity in India and 7 gigawatts of capacity in Brazil. The net saving involved would be $450 million for India and $930 million for Brazil ... Bizarrely, the utilities would still make major net savings if they simply gave CFLs away on street corners. The danger here is the old economic problem that if something is free nobody gives a hoot about throwing it away. A leasing system could be the solution.'

However, 'Energy Economist' comments, 'The fact that the potential savings are enormous does not mean that they will happen ... Yet there is much more at stake here than simple savings ... China will shortly be burning more than a billion tonnes of coal a year. India, too, is a massive coal user and only a few years ago had to shut down a power station because of emissions damage to the Taj Mahal. Is it really such a fanciful idea that development aid could be given in the form of CFLs?'

The question could be broadened to cover the whole range of modern, high-efficiency end-use energy hardware. The World Bank and other major funding agencies, as already noted, offer enormous sums of money in grants and low-interest loans to expand electricity supply capacity in Third World countries. The money would yield far greater returns in the form of energy services if a substantial fraction were devoted instead to installing high-efficiency lighting, motors, air-conditioners, refrigerators, freezers and other end-use hardware. To make this shift will entail, however, switching the financial benefits away from those who build power plants, towards those who manufacture the end-use hardware. Power-plant builders, already facing a steep decline in orders from industrial countries, will not readily yield such a considerable slice of their prospective Third World business; and their international influence is daunting. Their colleagues in the international funding agencies, still thinking of 'energy' as fuel and electricity supply, may be equally unprepared to redirect their attention to improving end-use efficiency.

Moreover, placing a single order for a gigawatt-sized baseload power plant, and administering its construction as a single project, can be done centrally, by officials in an air-conditioned office in New Delhi or Rio de Janeiro. Disseminating end-use hardware has to be done in the field, dealing with hundreds of thousands of individual energy-users all over the country, dealing with incomprehension, distrust and antagonism, often amid squalor, disease and wrenching poverty. Most planners instinctively prefer centralized policies that can be pursued in comfort –

even if they are demonstrably ineffectual, or indeed actively counter-productive. The problem is neither technical nor economic but political; but that does not make solving it any easier.

BACK TO BASICS

You stand on the pavement in front of the shop window, your nose pressed against the glass, gazing at the display inside. Your eyes dart from one fascinating item to another. Everything you can imagine desiring is there, a cornucopia of possibilities. You are spoilt for choice. Only one little nagging difficulty dulls your enthusiasm. Your pockets are empty.

In the 1990s, as we have seen in previous chapters, the earth's energy shop-window is bursting with opportunities. But most of the earth's people have empty pockets – if, indeed, they have pockets at all. For those in the rural Third World, the energy resource most desperately scarce is money. This alone means that energy issues in most of the Third World are almost unrecognizably different from issues familiar to those in the industrial world, or even in Third World cities.

To be sure, people the world over desire much the same energy services: cooked food, light, comfort, mechanical power, mobility, and so on. Indeed, throughout the Third World in the 1990s, the energy service often in most eager demand is, surprisingly, one of the most recent: that of electronics. As a result, ironically enough, even in remote rural villages in Africa, Asia and Latin America, radio and television now frequently display the global shop-window, including the extraordinary variety of energy services available elsewhere – for a price. In the rural Third World, however, the price is all too often far beyond reach.

No matter how daunting the energy problems faced in Third World cities, they are dwarfed by the energy problems of the rural areas. Most people on earth live far from cities, nomadically or in small settlements almost untouched by the industrial revolution. Imagine yourself one of this majority. You live as your parents, grandparents, and remote ancestors lived. You try to satisfy your basic needs for water, food and shelter as your forerunners did in bygone decades and centuries. Doing so, however, is becoming not easier but more difficult. You are almost entirely dependent on your immediate natural surroundings; somehow they do not seem able to sustain you as readily as they did in the past.

You may hunt food animals and gather food plants, or raise modest food-crops and keep domesticated animals like goats, sheep or cattle. Whatever your activities, they depend on daylight; nightfall means darkness, moonlight and starlight, assisted only by lamplight feeble at best. You find your water in natural waterways, or shallow wells dug with hand-implements. Its quantity and quality are unpredictable and deteriorating, especially since your sanitary arrangements are equally limited. You cook your food on open fires, burning wood you gather for the purpose. The energy you use comes from your own muscles, augmented occasionally by those of draught animals; from animal and vegetable oils and fats for lighting; and from virgin wood and charcoal for cooking. You use energy, in short, much as people used energy in the dawn of human history. You and three billion other people like you have yet to move much beyond the bonfire.

Yet unlike your ancestors, you and the other three billion living in the rural Third World often know that another more energy-abundant world already exists elsewhere. Where once you lived out your lives in unvarying isolation, now you may be visited by emissaries from the cities and even from other countries, variously offering to help improve your lot. You may encounter motorized transport like cars, tractors, boats, helicopters and planes; power equipment like drills, diggers and bulldozers; electric light; and electronics like radios and televisions, perhaps even microphones and video-cameras carried by film-crews. You and your fellow villagers may have a very clear idea indeed of how much easier your lives could be – if only you had the opportunity.

Until comparatively recently, however, energy emissaries from the industrial world plunged into rural Africa, Asia and Latin America like nineteenth-century missionaries, out to bring enlightenment to what they considered the 'heathen'. In the 1950s and 1960s, such forays into rural areas were usually undertaken to further one or another mega-project for fuel or electricity supply. The accompanying rhetoric invariably claimed that the project in question, for instance a massive hydroelectric dam, would bring better living conditions to the country involved. The project would be financed on generous terms, by either the World Bank or one of the regional development banks, or bilaterally between the Third World recipient country and a supplier country like the US or France. The financial support from the industrial side, however, meant that most of the cost of the project was paid more or less immediately into the coffers of the engineering and construction companies involved – almost invariably from an industrial country.

Moreover, although such energy-supply installations were often sited in rural areas, their output was delivered mainly if not exclusively to cities. The local impact was on balance more often negative than positive, defacing or flooding the land of rural communities, disrupting their social and cultural traditions and introducing new diseases. Even with the financial support of foreign aid, the output from a dam or power station still cost so much that rural communities could not afford to purchase either electricity or appliances to use it. Transmission and distribution systems usually bypassed local rural communities in favour of carrying electricity direct to urban areas.

This view of energy development in the Third World was closely akin to the traditional approach in industrial countries: the energy issue was to be resolved by expanding supplies of fuels and electricity. Planners paid little if any attention to what the fuels and electricity were actually for – what energy services they were supposed to provide. This traditional approach has proved inadequate even in industrial countries; in the rural Third World it verges on surrealistic. If you live in a modern industrial society, you take the role of money for granted, even if you do not have enough of it. You expect, for instance, to pay for energy services by some kind of financial means – cash, cheque, credit card. In the rural Third World, however, money is the scarcest resource of all. If you live in a remote village in India or Zimbabwe, you get your energy services, such as they are, by the sweat of your brow. Even if your government, perhaps supported by overseas aid, makes fuels and electricity available in your locality, you will be hard-pressed to find enough money to purchase suitable appliances, much less to pay for the cost of running them. This fundamental stumbling-block has long been overlooked by traditional planners dealing in broad-brush, hand-waving projections of Third World energy use.

COOKING IN A VILLAGE

Consider, for instance, the one energy service ubiquitous throughout the Third World: cooking food. Whether you live in Indonesian rain forest, Himalayan uplands or Patagonian pampas, you still expect to cook your food. Preparing food is an activity with endless variations from one culture to another; but cooking always requires a source of heat, and the commonest is an open fire. In physical terms, an open fire is an inefficient energy system; it may also have undesirable side-effects. If it is burning outdoors, most of the heat released from the fuel escapes immediately, rendering no service whatsoever. If it is burning indoors, the smoke and fumes make the room uncomfortable, to say the least, and discolour and darken everything in it. Traditional village cooking is basic. You put your Chinese rice or Mexican kidney beans or Zimbabwe sudza in a pot, probably earthenware, balance the pot on three stones over the fire, and wait. The inherent inefficiency of your energy hardware means that you may have to wait a considerable time.

Whether you light your fire indoors or out, you will almost certainly be burning mainly firewood, possibly augmented with dry animal dung or vegetable waste from preparing food. Throughout the rural Third World, firewood is the one essential fuel; and firewood supplies are rapidly dwindling. In recent years, in many different parts of the world, woodlands and other convenient sources of firewood close to settlements have been progressively depleted – in part by villagers themselves, but often by commercial logging or more drastic commercial clearing. In many cultures, women not only do the cooking but also gather the firewood. To do so they now often have to walk for several hours just to reach a usable supply of wood, and several more hours to return carrying such firewood as they have been able to find. A village woman may have to make a foray like this every day; the journeys are getting longer, and the firewood ever scarcer. Furthermore, the denuded landscape, stripped of its tree cover, loses water more rapidly in dry seasons, and cannot regain enough of it in wet seasons. Rain and wind scour the unprotected surface of the earth, reducing its fertility in a vicious spiral of deterioration. Fragile landscapes can degenerate with appalling swiftness into deserts.

One possible remedy might be to use firewood more efficiently. Even in the 1950s, aid and development agencies working in rural Africa, Asia and Latin America had begun seeking ways to help villagers improve their cooking arrangements. At the time, agency fieldworkers were only incidentally seeking to reduce firewood requirements by using it more efficiently. They were more concerned to lighten the burden on village women: to reduce the time they had to spend preparing food; to minimize burns and other kitchen accidents; and to clear the indoor atmosphere of smoke and fumes. Replacing traditional cooking fires with better cook-stoves could further all these objectives, and use firewood more efficiently as well. Doing so, however, proved much more difficult than well-meaning outsiders initially anticipated. Indeed, for a time projects to upgrade village cook-stoves, however well-intentioned, looked all too often almost as heavy-handed and ill-thought-out as programmes for massive power-supply installations.

Fieldworkers introduced stove projects to villages without adequately investigating actual village customs and cultures, including cooking practices and social conventions. 'Improved' stoves, designed and tested under laboratory conditions in urban institutes, performed poorly in village conditions. Villagers did not have the means to maintain them, or did not know how. Pots did not fit properly on top. Wood could not be inserted without cutting into smaller pieces. Chimneys leaked or plugged; and so on. Many of the stoves distributed in the early projects fell into disuse virtually as soon as the fieldworkers departed. Moreover, the early designs cost too much to fabricate. Without major subsidies from aid agencies or central governments few villagers could afford them.

On through the 1980s, however, the aid agencies persevered, learning the lessons from unsuccessful stove projects and refining subsequent projects accordingly. Most of the initiative came from non-governmental organizations, like the Intermediate Technology Development Group, which had been founded by E.F. Schumacher in 1965, and similar groups, and by universities and institutes in the countries themselves. As stove projects began to establish themselves, governments too joined in. They sent officials to tour villages and explain arrangements. They fostered training schemes for stove-making. They sponsored backup services for trouble-shooting and maintenance. Progress, while frequently obvious and encouraging, was frustratingly slow; but by the 1990s some projects at least were beginning to take off.

Analytical studies, like the 'Household Energy Handbook' prepared by Gerald Leach and Marcia Gowen for the World Bank, published in 1987, revealed that a successful stove project requires scrupulous attention to local detail of every imaginable kind. How do local people in the highlands of Chile or Nepal, or the plains of Zimbabwe or India, or the forests of Indonesia or Brazil, use their fires? Do they use them only for cooking, or for heating water for drinks or washing, or for indoor comfort, or as a focus for social gathering? What do they cook – rice, beans, other foods? When? In what vessels – clay, terracotta, cast iron, aluminium? How long does it take? Should the proposed innovation be based on a stove made of mud, or ceramic, or metal, or a combination of materials? Should it take one pot, or two, or even more? Should it have a chimney? Who should make the stove, and how? The householder? A village artisan? An artisan in the nearest urban centre? Who is to provide the materials? What will each stove cost? Who will pay for it, and how? How long will it last? What maintenance will it require, and by whom? Every specific locality has its own attributes, its own idiosyncrasies; each must be approached on its own terms, if outsiders – even from the nearest city – are to render genuine help.

Traditional energy thinking, based on aggregates and averages, is even more inappropriate in the Third World than it is in industrial countries. In the rural Third World it is not merely misleading but actively pernicious. Disrupting the already precarious life of a remote village by introducing putative 'improvements' that do not work is no kindness. The Intermediate Technology Development Group, one of the pioneers of improved stoves, surveying the situation in 'The Stove Project Manual', published in 1985, was uncompromising about the failure of many projects to achieve their objectives.

In the 1990s, nevertheless, despite numerous disappointments and stubborn difficulties, some projects are beginning to make a genuine difference to life in rural villages, and also to the life of the urban poor, in countries of widely differing geography and culture. The programme led by the Indonesian group Dian-Desa has disseminated over 25 000 ceramic stoves; the programme led by the Kenyan group KENGO over 180 000 'Jiko' stoves; and the programme led by the Indian group ASTRA over 180 000 mud stoves. The 'Chula' mud stove developed by the NADA group led by Madhu Sarin in northern India has also achieved encouraging acceptance. Other active programmes include those in Guatemala with its 'Lorena' stoves, the Gambia, Nepal, Sri Lanka and Zimbabwe, among others. To be sure, by no means all the stoves disseminated are still in use. Some individual stoves do not work properly; some wear out or break; some are simply abandoned by village women who prefer the old ways. But many stoves have been adopted enthusiastically, becoming a source of pride and satisfaction to those who use them – and perhaps even a harbinger of further opportunities to improve their lives.

In the village of Dema, in Zimbabwe, for instance, the village women pooled their efforts to learn to build improved stoves, working together. Their kitchens are now much cleaner; the children cannot pull the pots on top of themselves; a bundle of firewood lasts a week; and the women celebrate their achievements in exuberant communal song: 'Say what you want! What about the hands? The hands are ours! What about the energy to do things? The energy is ours! The brains are ours! . . . The kitchen is ours! . . . The country is ours! . .. The action is ours! . . .' In the village of Murewa a woman seated by her new stove says with shy delight 'It makes you feel young again!'

A start has been made; but funds, even for basic stove materials, are limited, and trained fieldworkers far too thin on the ground. Moreover the truly poor cannot afford to purchase even a simple grate, much less an improved stove. Only a serious commitment, not only by Third World governments but also by industrial governments and international funding agencies, can accelerate stove programmes enough to make a real difference.

Successful stove projects underline one parallel between the industrial world and the most remote corners of the rural Third World. A successful stove project can reduce firewood requirements much more significantly and rapidly than supplies of firewood can be expanded. From midtown Manhattan to outer Mongolia, therefore, the first priority is to provide the end-use energy service as efficiently as possible. In the Third World, improved stoves must be part of the answer. The funds are available; but those who disburse them must shake off their fixation with traditional supply-oriented policies, and redirect attention, money and qualified people to those who so desperately need improved end-use energy services – even in the unglamorous low-tech form of better cookstoves.

TREES

Trees are the original fuel, solid solar energy stored for use. Most industrial countries, however, have long since destroyed their great forests; and wood has long been superseded by fossil fuels. Only recently have energy analysts in industrial countries like Sweden begun to look upon wood again as an important fuel resource for the future. Meanwhile, Third World planners, taking their lead from the industrial model, have likewise concentrated their attention on 'modern' energy carriers – especially fossil fuels and hydroelectricity – even as their own indigenous resources of fuelwood are being remorselessly eroded.

In principle, fuelwood is a 'renewable' resource; but only if new trees are growing to take the place of those cut down. All too often, however, commercial forestry in Third World countries is an 'extractive' industry reminiscent of open cast mining, and with an impact just as drastic. Vast areas of mature trees are 'clear-cut', leaving nothing standing, not even saplings. Shorn of its tree-cover, the land may be rapidly drained of its nutrients and even of its topsoil. Any fresh growth that does emerge is likely to be shallow coarse scrub, a stark contrast to the rich vegetation of the original forest. No amount of reforestation can hope to offset the ravages of logging like this.

On a global scale, compared to such depredations, the activities of village women scavenging for firewood are trifling. In a particular neighbourhood, however, the need to cook enough to feed more and more mouths can deplete the local forest all too swiftly. One obvious remedy is to plant more trees. In most rural Third World areas, trees are just a fact of nature, there to be gathered for firewood – or, more and more frequently, not there. The notion of trees as a 'crop', to be planted, nurtured and eventually harvested, is unfamiliar.

Initially, policies aimed at replacing lost trees, like so many other policies, were devised and administered centrally. The foresters worked on tidy plantations; they had no contact with people in the villages, and the plantations were fenced and guarded. According to Yemi Katerere, director of the Zimbabwe Forestry Commission, for instance, 'The evolution of forestry in most former colonies was based on the scientific management of forest. This implied that you had to set aside large tracts of forest and you had to keep the people out. The people reacted to this: they saw us foresters as enemies. "These people have excluded us; we don't have access to this resource, and we might as well set it alight". Now we have accepted that forestry needs to be integrated into the fabric of social life of Zimbabwe. We need to manage the forests for the people. More and more, instead of spending resources on policing, we spend these resources on educating the people.'

One corollary of this changing view is that tree-planting and nurture are also being integrated into local societies and cultures, rather than instigated and administered from some distant city. The Green Belt movement of Kenyan women, inspired by Wangari Maathai, is a remarkable demonstration of activity at the grass-roots, or rather at the tree-roots. The Green Belt women gather and plant tree-seeds by the thousands, watch over tree-nurseries, plant out, water and feed the saplings, and spread the gospel of reforestation with exhilarating commitment. Zimbabwe, too, has tree-parties with songs and dances, as the idea of actually planting and caring for the coming generation of trees catches the imagination of rural people.

The value of trees – stabilizing local ecology, retaining water, providing shade, sometimes yielding fruits or nuts, and eventually becoming essential fuel – is ever more widely understood, even in remote areas. Troubling questions nevertheless intrude. Who owns the trees, or the land they grow on? Who is responsible for looking after them, watering them, protecting them from cattle and goats? Who is entitled to cut them down – and who is not? What happens to an impoverished transgressor who presumes to cut wood off someone else's tree? What if a landlord denies trees to local villagers in order to sell the firewood in the nearest city? Even trees, the worldwide symbol of unsullied nature, are becoming political.

LIGHTING

'Let there be light.' Those of us who live in industrial societies can no longer feel the original visceral impact of this dramatic statement. We can mimic the Old Testament Deity merely by flicking a switch. For most of the earth's people, however, sunrise is still a daily miracle; daylight is the brightest and most reliable light they have. After sunset, the poorest have to rely on the glow from their cooking fire. Some, more fortunate, have candles or kerosene lamps, or even an electric torch, powered by batteries and used only when absolutely necessary. Some even have electricity; but its supply may be limited and unreliable. For all of them, sunrise is the day's most important event.

Accordingly, although cooking is undoubtedly the most essential energy service people desire, after cooking comes better illumination. If you have light after sunset, you can extend your working day. A weaver can make more cloth, an artisan more pots or baskets – or indeed cookstoves; a farmer can maintain tools. Above all, a villager whose daylight hours must be spent in essential physical toil may still have an opportunity for education, reading and study – crucial for those desiring better lives, and especially the young.

A kerosene lamp, however, may be only a wick in a bottle, giving light as feeble as a candle and smelling worse. A pressurized kerosene or benzine lamp with a mantle overcompensates in the opposite direction, giving a garish glare reminiscent of a nineteenth-century arc light, and just as noisy. What was true a century ago in industrial countries is now equally true even in remote rural villages: the key to better illumination is electric light – and electric light in turn requires electricity. In a remote rural village, unfortunately, saying 'Let there be electricity' is likely to be as ineffective as saying 'Let there be light' at midnight. Cooking, even on an innovative and effective stove, can still depend on burning firewood whose cost is paid in personal exertion, not cash. Electric light and the electricity to run it, however, cost money – sometimes a great deal of money, and in any case more money than most rural villagers can afford.

British analyst Gerald Foley surveyed the dimensions of the problem in a wide-ranging report entitled 'Electricity for Rural People', published by the international development organization Panos in 1989. According to Foley, planners have two options: to extend the interconnected electricity grid ever farther, from cities into outlying areas; or to establish isolated local generating and distribution systems. The first option has significant technical advantages. Generating plant can be larger and more efficient, costing less per unit of output; fuel and spare parts are easier to deliver; trained staff can be readily available to handle routine maintenance and breakdowns.

Unfortunately, however, as the grid expands from densely populated cities toward a rural population scattered over a vast area, the cost of installing and maintaining transmission lines shoots up rapidly. Taking a line of suitable voltage to a small settlement of a few houses is punishingly expensive, especially when those who live there can ill afford to pay even for a single lightbulb, much less for the electricity to run it. Planners must then decide which villages – and indeed which particular villagers – are to receive the necessary major subsidies, not only for connection but also for bulbs, other appliances and running costs. Central authorities may subsequently claim that 'all the villages in a certain region now have electricity'. The claim may sound impressive politically; but its practical meaning may be that one or two houses in each village are connected, and may not have much more than a single lightbulb apiece. As a way to use capital and skills already in desperately short supply, such a programme falls lamentably short.

The alternative is self-contained electricity supplies, isolated and unconnected to a grid. The system might be based, for instance, on a diesel or petrol generator connected directly to nearby premises. This arrangement eliminates costly long transmission lines, and has the potential advantage of giving local people themselves control over their electricity supply. However, even transporting a generator of any size over rudimentary roads to a remote location may itself be both expensive and brutally difficult, and delivering fuel to it almost equally so. Experience suggests that obtaining spare parts will be a perennial headache; and few villagers have the requisite know-how to cope with either normal maintenance or breakdowns. With no grid connection, a breakdown of a local generator means no electricity, almost certainly for some days and possibly indefinitely.

The local generator need not, of course, have a diesel or petrol engine. In Nepal, for instance, many small watermills, originally used to supply direct mechanical power for grinding grain, have been fitted with generators to supply electricity; numerous small hydro installations are also operating in northern Pakistan. Another possibility is an array of photovoltaic (PV) solar cells. As noted in the preceding chapter, PV arrays are still too costly to compete with traditional supply technologies in industrial countries. They are, however, competitive with isolated diesel or petrol generators in remote locations – provided the users can afford either option.

According to Bernard McNelis, in his report 'Solar Electricity: a survey of photovoltaic power in developing countries', published by Intermediate Technology and UNESCO in 1988, several thousand PV systems have been installed in developing countries, mostly since the early 1980s. 'The size of these systems ranges from a few watts to over 30 peak kilowatts, for applications as diverse as water pumping, vaccine refrigeration, domestic lighting, cattle fencing and telecommunications. Many of the systems have been of an experimental nature, for developing and demonstrating the technology, but increasingly photovoltaic systems are being installed for sound commercial reasons, as being the most cost-effective solution for particular applications.'

Several thousand small PV systems for domestic lighting for individual premises have been installed, particularly in the South Pacific, French Polynesia and China. According to McNelis, 'They are simple to operate and reliable, now that earlier problems with battery charge regulators have been solved. PV-powered fluorescent tubes provide a much higher quality of light compared with candles or kerosene wick or pressure lamps.' However, using PV arrays for more ambitious rural electrification schemes for entire villages presents problems. 'Central systems of this type are vulnerable to failure due to component faults or over-loading and thus need a high level of supervision.' Moreover, if the PV array is to operate not only lighting but more demanding hardware like motors and electronics, the electricity from the array has to be put through an 'inverter', to give it the correct voltage and frequency. Such inverters, says McNelis, 'have not proved particularly efficient or reliable in the past'.

The Marymount Hospital in Zimbabwe, for example, installed a large PV array in 1982. A sign outside the hospital proclaims the array 'A Gift from France'. The hospital staff disposed of two of their three diesel generators, keeping the third only as a standby. Unfortunately, a month after the PV array was installed it broke down, apparently because of the inverters. Since that time the hospital's one doctor has had to rely on the remaining diesel generator. Its capacity is so limited that much of the hospital's expensive electrical hardware has to be left unused, and the rest switched on and off so that only the most essential is operating.

In 'Electricity for Rural People' Gerald Foley comments bluntly 'It is sometimes argued that without extensive practical experience and a reasonable market for manufacturers, renewable technologies will never become technically and commercially viable. This is true. It does not necessarily mean that the villages of the Third World should be used as a test-bed for the development of technologies which will primarily benefit manufacturers in the industrial world. There may, indeed, be cases where such projects can usefully take place with the informed consent of all concerned; but it is important that this kind of commercial product-development should not be disguised as aid. Where experimental projects are carried out, the welfare of the local people involved should be a paramount concern. If equipment is provided, there should be provision for its long-term maintenance and replacement after the project has ended. It is not acceptable that people should have their lives disrupted and expectations raised by projects which temporarily increase their standards of living but leave them with equipment which they can neither maintain nor replace.'

Foley's report lists more than two pages of concise conclusions. The last are categorical: 'There is no competition between the conventional approach to rural electrification and the promotion of locally-managed off-grid systems. Both are needed and both should be supported. The final conclusion is positive and emphatic. Well-chosen rural electrification programmes are worthy and attractive options for donor funding. There are strong reasons for believing that the amounts of money devoted to them will significantly increase in the future.' If the rural Third World is to have any hope of catching up with the rest of humanity, Foley had better be right.

LIGHTING AND BIOFUELS

Firewood, for all its virtues of simplicity, is not much use for lighting – not, at least, in its original form. It need not, however, be left in this form. Like coal, wood can be 'gasified' to produce a fuel gas, entirely suitable for use in gas-lights exactly like those used in industrial countries a century ago, and still used for camping. As described in the previous chapter, the principle of the gasifier is simple and straightforward. Wood of suitable size is burned in a sealed container with a restricted supply of air, possibly with added water, although the wood itself usually contains enough water to sustain the desired chemical reactions. The process produces a mixture of carbon monoxide and hydrogen, almost identical to the 'town gas' used for nearly two centuries in the industrial North. This fuel gas can be piped from the gasifier direct to gas-lamps with jets and mantles, to give bright and easily controllable light. The fuel gas can also be burned in a cooker with a higher efficiency than burning the wood in an open stove, or even an internal combustion engine, to run a pump or other mechanical hardware.

As noted in the previous chapter, 'biomass gasification' on a large scale is already an established technology in countries like Sweden and Finland. Its output, powering gas turbines in combined cycle plants, offers a way to make maximum use of the fuel-energy in Scandinavia's forests; and the same technology could do likewise for forested areas of the Third World, especially in Africa, Latin America and southeast Asia. Instead of a generator requiring diesel oil or petrol brought in over difficult terrain, a local electricity system could use a gasifier and a small gas-turbine, fuelled with wood and wood waste from the neighbouring forests. To be sure, all the provisos and qualifications itemized above would still apply. The capital cost of even a single unit, including not just a gasifier and gas turbine-generator, but also heavy-duty hardware to pulverize the biofuel into suitable form, would be substantial. Getting the unit to the site would be a major task. The plant would have to be supervised by qualified staff; and spare parts would have to be readily available. Nevertheless, in the longer term, in regions with the appropriate resources, biomass combined-cycle plants could well become a major contributor to electricity supply. In the short term, even small-scale gasifiers for local lighting will have to clear the usual hurdle of cost.

Like firewood, unfortunately, dung is not much use for lighting in its original form. Like firewood, fortunately, dung can be readily converted into a much more suitable biofuel, as described in the previous chapter. If you stir dung, which may include both animal and human waste, into a thick liquid slurry and let it ferment in the absence of air, bacteria in the slurry give off 'biogas' – largely methane, closely akin to natural gas. If you ferment the slurry in a capped container or 'digester', you can pipe the methane directly to gas jets in lamps or cookers, or even to an internal combustion engine like a pump. When the slurry has yielded enough methane, you can drain it into a drying area; the remaining solid is even better fertilizer than the original dung and much easier on the nose.

Such 'biogas plants' have been used for many years. China is said to have hundreds of thousands, if not millions, and they have also been installed on large farms in many parts of the industrial world. More recently they have also become common in the Third World, both as communal facilities for entire villages and also to supply individual households. Anyone with perhaps four cattle can collect enough dung to keep gas-lights burning, and perhaps also to power a gas cooker or even a gas-fired pump.

Kalyanpura, for example, is a 'demonstration village' in Gujarat in India, in which the Gujarat Energy Development Agency has installed a wide range of innovative energy technologies. Kalyanpura's village-scale biogas facility entails gathering cartloads of dung from all the participating villagers, for delivery to the biogas plant. The work is unpleasant manual labour, carried out by Harijan 'untouchables' for minimal pay, an aspect of biogas use that tends to be overlooked. But the biogas powers streetlights, house lights and cookers, and has dramatically improved the living standards of those who use it. The spent slurry from the plant is dried and spread on nearby farmland, markedly improving its yield.

In rural Zimbabwe hundreds of individual homes have simple Chinese-style biogas plants with brick or concrete linings sunk in the ground; more would be built if supplies of the requisite cement were not so hard to come by. Almost anywhere you look in the Third World, using even such a simple and effective technology as a biogas plant is still impeded by scarcities of the most basic materials.

TRANSPORT

Among all the energy services the industrial world takes for granted, none poses such problems in the Third World as transport. Third World cities, following the industrial model, are now choked with cars, many of them elderly, inefficient and polluting. Even in the urban shantytowns, dilapidated hovels have cars standing outside. Yet the majority in the rural Third World must rely on their own muscles and those of animals – oxen, donkeys, water-buffalo, llamas, any living creature that can be domesticated and fed – to move not only themselves but also the heaviest burdens. Roads are often barely discernible, frequently impassable for motor vehicles, and in many areas non-existent. Localities close to cities may be served by buses, and here and there railways, whose rolling stock is usually spartan if not decrepit. In much of the Third World, however, moving people and goods is still in the Dark Ages.

Forget for a moment the staggering scale of investment in roads and vehicles that would be required to make a major difference. What if, somehow, people in the Third World used cars in proportions comparable even to those in Europe, to say nothing of North America? The outpouring of carbon dioxide alone would utterly overwhelm any other measures that might be taken to mitigate the greenhouse effect, to say nothing of the environmental impact of other exhaust gases. Nevertheless, as cities throughout the industrial world mount last-ditch struggles to rescue themselves from the stranglehold of the car, Third World planners are eager to plunge their citizens into the same quandary.

No easy answer comes to mind. Across the entire planet, transport, with its requirement for portable fuels, is the most intractable long-term energy problem of all. Thus far it has received too little attention even in the industrial world. Henceforth it must rise to the top of the agenda, before the Third World makes all the same mistakes, and the problem becomes comprehensively insoluble.

PRESENT IMPOSSIBLE

Ironically, one consequence of improved transport services has been the exodus of the young and able from the rural Third World into the cities. Perhaps the most immediately critical challenge to Third World planners is the urgent need to upgrade amenities in rural areas enough to persuade at least a core of skilled, educated people to live there – not just to visit, but to reside. Unless doctors, teachers and technicians are willing to remain in remote areas, using their abilities to help rural people improve their lot, the plight of the deprived can only grow steadily worse. Moreover, the deterioration of conditions in rural areas will drive more people off the land and aggravate already serious problems in the overcrowded cities. Basic energy services, for lighting, cooking, and pumping, for clean water and sanitation, will be crucially important to keep a core of skilled people in remote villages.

Centralized planning on the traditional model, with its emphasis on major supply projects like dams and power stations, simply cannot meet the need. The World Resources Institute report 'Energy for a Sustainable World' argues that Third World development should take the alternative route. Third World countries have one signal advantage: they do not yet have vast countrywide infrastructures of inefficient buildings, industrial processes and transport systems, like those which impede change in the industrial world. Instead of following the traditional path of the industrial countries, Third World countries can 'leap-frog' over the inefficient stage, to take advantage of the latest understanding about energy and energy services. 'Energy for a Sustainable World' argues that the balance of effort, investment and training must be shifted, from the traditional approach based on centralized large-scale development of fuel and electricity supplies, to the alternative approach, based on providing decentralized end-use energy services with the highest possible efficiency. Only thus can the Third World marshal limited resources to obtain the maximum benefit for the most people. With the appropriate support, end-use hardware of the highest efficiencies, like compact fluorescent lights and electronics, especially telecommunications, powered by decentralized technologies like small hydro, gasifiers, biogas plants and photovoltaic systems, can make life in rural areas less of an ordeal, not only for those born there but also for the fieldworkers helping them to improve their circumstances.

Nanubhai Amin, founder of the Gujarat Energy Development Agency in India, points out that traditional supply technologies have taken many decades to establish themselves. He notes that the approach based on end-use efficiency and renewable energy has only been pursued for some fifteen years; it may need forty years to come into its own. The same applies right across the Third World. To succeed even on that timescale, however, it will require a profound reassessment of the policies of international funding agencies and industrial governments. Time is short. If it is not already too late, it may soon be.

Amulya Reddy, the India co-author of 'Energy for a Sustainable World', puts the position succinctly. 'It is going to be extremely difficult to implement the measures we propose – but not more difficult than continuing with the present, which is just impossible.'

10
FUTURE ALTERNATIVE?

The earth is in trouble. So are its people. We have come a long way, many of us, since that first intentional fire, millenia past. But we have left most of our fellow human beings behind; and the energy traditions of our industrial world may be leading us all up a blind and dangerous alley.

What do these traditions tell us now? In September 1989 the World Energy Conference met yet again for its triennial convocation, this time in Montreal, Canada. It concluded that world fuel and electricity use would increase as much as 75 per cent by 2010. Two-thirds of this increase would come from fossil fuels; nuclear electricity output would more than double. Far from decreasing in line with the urgent call from the Toronto conference the previous year, carbon dioxide emissions would increase by 70 per cent.

In the same month, the European Commission published a study entitled 'Major Themes in Energy to 2010'. Its first scenario, a 'conventional view of the future', concluded that fuel and electricity use within the European Community would increase by 37 per cent by 2010, and carbon dioxide emissions by 24 per cent. Two further scenarios wrestled with the implications of this conclusion, without finding much cheer. The third scenario suggested that a further 112 large nuclear stations be completed and in operation within the Community in the next two decades, more than doubling existing nuclear capacity. Confidential projections by Britain's Department of Energy, in a document 'leaked' to the Association for Conservation of Energy, concluded that Britain could be producing 34 per cent more carbon dioxide by 2005, and 75 per cent more by 2020.

Wherever we look, the energy tradition tells us that we must continue using ever more fuels and electricity, essentially without limit. We must therefore mine and burn ever more coal; extract and burn ever more oil and natural gas; mine, process and use ever more uranium. We must build ever more coal terminals, pipelines, refineries, power stations, and other supply facilities, at a capital cost of billions upon billions upon billions. Yet the majority of people – those in the rural Third World –

are doomed to lag ever farther behind. No matter how far forward the official scenarios point, their graphs are still climbing relentlessly, still vanishing out the upper right-hand corner. Even though they see us using twice as much fuel and electricity two or three decades hence, our voracious appetites will not have been sated even then. We shall still want more, and ever more. We shall still insist on ripping ever more resources out of the earth, pouring ever more carbon dioxide into the sky, accumulating ever more plutonium to keep track of, as the gap between rich and poor gapes ever wider.

As a vision of our global future, it is bleak, and growing bleaker. Is this really the best we can do? More and more people do not believe it is. More and more people are now seeking a 'sustainable' way of living on this planet – a way of living that respects the fragile balances of the biosphere, yet offers all of humanity a richer life. Can we provide the energy services to enable us – all of us – to get there from here? If we accept the traditional official forecasts, projections and scenarios, and the assumptions that underlie them, the answer is simply 'no'. According to the energy tradition, we are trapped in a rut that must lead us inevitably to catastrophe.

We need not and must not accept this fatalistic prognosis. There is an alternative. We know we can do better; the opportunities abound. But the energy alternative is not just a technological shopping list. It is a new way to think about energy. The energy tradition is preoccupied with supplying fuels and electricity, to meet an extrapolated, aggregated 'demand' presumed to arise almost willy-nilly. The energy alternative starts from a fundamentally different premise – one that is so obvious it is far too easily ignored: no one wants 'energy', nor yet fuels and electricity. We want energy services. If we really want to reduce the environmental impact of using fuels and electricity, we can readily do so. The better the energy hardware, the less fuel or electricity we need to run it. By improving the end-use hardware, we can reduce dramatically the environmental side-effects of using energy.

EFFICIENCY IN ACTION

We already know how to make energy hardware much better than traditional designs, as we have seen; and we are still learning rapidly. Existing hardware with a long service life, like buildings, we can upgrade to enhance performance. New hardware with a shorter life service life, like appliances, industrial plant and vehicles, we can manufacture to more exacting designs. We can retire and replace old inefficient hardware with the best available. We can encourage the adoption of these measures by persuading governments to legislate more stringent criteria for end-use hardware. The most enlightened governments have already done so, and others are following suit.

Building regulations, for instance, can stipulate high minimum standards of insulation. The thermal performance of a house can be measured and recorded in legal documents; building societies and other mortgage agencies offering financial support for house purchases can offer preferential terms for better buildings, and require upgrades for inadequate ones. Architects can take advantage of the rapidly expanding body of experience called 'passive solar' design, making the best possible use of sunlight at any particular location, to provide warmth and light indoors. Governments can insist that the efficiency of domestic appliances be tested and stated explicitly on labels, to enable purchasers to choose a new cooker or deep-freeze with the most complete possible information about what it may cost to run. Governments can also ban the sale of appliances that do not meet minimum standards of efficiency. Governments, as major purchasers of energy hardware, can specify high efficiency as a guideline for their own purchasing policy, stimulating and expanding markets for everything from insulation to compact fluorescent lights to fuel-efficient vehicles.

Governments have long provided generous financial support to supplying fuels and electricity; they still do so, nationally and through international agencies. They offer grants, tax relief, low-interest long-term direct government loans, and government guarantees for loans from private financiers. They finance research, development and demonstration of new supply technologies, especially nuclear technologies, and even offer insurance cover for allegedly commercial nuclear installations. Given the necessary political will, a substantial proportion of these financial resources can be redirected into improving end-use energy hardware. Some governments, to be sure, have made token efforts in this direction, like the limited grants offered to householders in Britain to insulate lofts. But a genuine government commitment to end-use efficiency will require financial backing on a scale far greater, and an administrative programme perhaps akin to that mounted in the 1960s, to convert the whole of Britain's end-use hardware from town gas to North Sea natural gas.

The precedents are there; thus far the commitment is not. Nor can such a programme be mounted without vociferous and potent opposition from at least some suppliers of fuels and electricity. Their influence in industrial society is equalled only by that of the weapons-producers. The fuel and electricity suppliers have enormous coffers, immediate access to the inner sanctum of government, and a central role in the economic life of every industrial country. In the past two decades they have defined the terms of the energy issue in their own interest – not through malfeasance but merely through habit and tradition. Redefining the energy issue to cope with the true reality may well entail redefining the role of the fuel and electricity suppliers. The objective can be to reshape them into the role that Edison originally envisaged, to supply not fuel or electricity but the energy services people actually want. Many US utilities and their regulatory overseers are already fully engaged in this process, directing new investment not into yet more power stations or gas pipelines but into more efficient buildings, appliances and industrial plant. 'Least-cost planning' demonstrates that the approach has clear-cut advantages not only for the environment but also for the balance-sheet. As enlightened fuel and electricity suppliers recognize where their own interests lie, the transition from tradition to alternative can only accelerate. Governments and regulators can hasten the process by amending fiscal, legal and licensing arrangements as appropriate. Examples already abound.

ENERGY CARRIERS

However efficient we make our end-use hardware, we shall still need energy carriers to run it. As the thermal performance of buildings improves, we shall need much less low-temperature heat for comfort. But we shall still need higher temperature heat for cooking and hot water, and for industrial processes. We shall need electricity for lighting, motors, electronics and electrochemistry; and we shall need portable fuels for transport. The traditional energy carriers – coal, oil, natural gas and electricity – each present both problems and opportunities, as we have seen. All will remain important for longer than anyone can foresee; but their relative importance, and their practical roles, may evolve dramatically.

We can get heat from many different sources – not only the traditional energy carriers, but also renewable sources like sunlight, wind and biomass. We can of course supply the heat directly where we want it, for instance by delivering a traditional energy carrier to the premises and using it there. We can also, however, integrate heat supply on a larger scale. Ordinary water can act as an energy carrier for heat. With suitable plumbing we can circulate it not only through an individual building but throughout an entire city, gathering heat from all available sources – people, lights, appliances, computers, industrial processes, burning fuels, sunlight – and delivering the heat where it is desired.

Supplying heat is easy. Supplying electricity is somewhat more difficult, but increasingly important. Improving the efficiency of end-use hardware will reduce the total amount of energy carriers we use; but it will also make electricity a progressively larger proportion of this smaller total, because of electricity's distinctive properties. Generating electricity can be much more efficient and environmentally acceptable, as we have seen. The traditional approach to electricity supply, emphasizing enormous centralized base-load power plants, is giving way to smaller plants, more numerous and diverse. Smaller power plants can be sited more easily, and completed more rapidly. They require less space and less water, and need not dominate their surroundings or intrude too much on amenities. They can achieve much higher efficiencies in operating modes like combined heat and power – cogeneration of heat and electricity – and combined cycles; and they can incorporate innovative technologies like fluidized bed combustion and coal gasification to reduce noxious emissions. A system based on widespread diverse sources can accept electricity generated by small hydro, wind power, wave power, tidal power where available, and photovoltaics, as they become technically and economically established. The more electricity generated from renewable sources, the less that need come from fossil or nuclear fuels, with their inevitable environmental impacts on air quality, the greenhouse effect and toxic waste disposal.

Of all the energy services, the one that presents the most intractable problem is transport, especially the car. The problem arises not merely from the car alone, but from the effect it has had on the structure of our societies. We can make the internal combustion engine more efficient; we can reduce its noxious emissions; and we can supply appropriate portable fuels including not only petroleum but also compressed or liquefied natural gas, methanol or ethanol. We can even introduce electric cars. But we have laid out our cities to depend ever more on cars, to carry us from home to work to shopping to leisure and back again; and no matter how much road space we pave, the cars proliferate to jam it.

The energy service of transport is already a major problem, but not an energy problem alone. To be sure, even when cars have catalytic converters to minimize other pollutants, traditional portable fuels like petrol and diesel are relentlessly increasing the carbon dioxide concentration in the atmosphere. A fuel like ethanol, based on biomass, would alleviate this problem. But in Rio de Janeiro, where most cars burn ethanol, traffic is grinding to a standstill just as it is in so many other places. More than any other energy service, transport has degenerated pathologically. Coping with the transport problem will pose more difficulties than all the other energy services combined; and the prognosis is not good.

THE THIRD WORLD DILEMMA

The transport problem in the Third World is fully as daunting as that in the industrial world, albeit for the opposite reason. The industrial world has too much concrete and steel already in place, in the wrong places. The Third World has too little. This represents both a challenge and an opportunity, not only for transport but for all the uses of energy. The challenge is obvious: Third World countries must somehow find the resources – money, technology and skills – to create systems to provide essential energy services to all their citizens. The opportunity is less obvious: Third World countries can do it right the first time. Instead of burdening themselves with an array of energy hardware both inefficient and polluting, they can 'leapfrog' over the inefficient, polluting stage direct to the best available designs. But they cannot do so on their own; they will need all the help they can get.

Unfortunately, the help they now get comes largely from governments, corporations and international agencies steeped in the energy tradition. They focus narrowly on supplying fuels and electricity, instead of supplying the services people actually require. Such policies can only accelerate environmental degradation, and widen the chasm between 'haves' and 'have nots'. They can and must be changed, to focus on meeting the basic human needs of the most deprived, where they arise and as they arise, to help them to help themselves. As the previous chapter indicated, only this alternative approach has any hope of succeeding.

A SUSTAINABLE WORLD

The World Resources Institute report describes 'Energy for a Sustainable World'. In the long term, what might such an energy system be like? In principle, it could rely on just two energy carriers: biomass for fuel – solid, liquid and gaseous – and photovoltaics for electricity. The solar energy falling on the earth could readily supply both. No fossil fuels would mean no excess carbon dioxide added to the atmosphere. As innovations continue, we could find that industrial processes came to rely primarily on ambient temperature technologies with high efficiencies. Amory Lovins offers a provocative example: 'We know three ways to turn limestone into building material. We can cut it into blocks; we can roast it at a high temperature to make cement; or we can feed it to a chicken.' Weight for weight, eggshell is one of the strongest building materials known; but we do not yet know how the chicken does it. The process is remarkable not only because its product is so impressive, but because the process itself takes place at the chicken's body temperature, not much above ambient. Such a process loses very little energy as wasted heat, and is accordingly very efficient. Ambient temperature technologies, like biotechnology and solid-state electronics, could become progressively more important. Instead of changing the world by using energy with brute force, we could change it with precise and subtle elegance.

All that, however, lies far in the future. We cannot plan for the twenty-second century, nor do we need to. We only discovered petroleum as a valuable fuel less than 150 years ago. What matters is to choose the right priorities today, to widen our options. If we get the next decade right, the twenty-first century will look much more promising. The encouraging developments outlined above can all happen. But will they? Governments and corporations are still trammelled by the energy tradition; they will not act on their own. We – especially those of us in industrial countries – must apply urgent, insistent pressure. The environmental constraints on energy use are ever more pressing; but society's institutional arrangements are impeding desirable changes. Energy decisions must not be left to those who still think 'energy' equals fuels and electricity.

Can the energy tradition overcome its stagnation, and evolve rapidly enough? Can the energy alternative cope with its daunting economic and social uncertainties, and achieve its apparent potential? Or will our planetary systems overload, break down and collapse, taking humanity down with them? Within our lifetimes we shall probably find out, for good or ill. The earth is in trouble; so are we. But we can change the way the world works. Will we?

Strike a match, and think about it.

APPENDIX A

'ENERGESE': the language of energy policy

Every time you encounter one of the following words or phrases, beware. What does it really mean? What assumptions does it carry with it, unstated? This glossary of 'energese', with some of its possible meanings, is far from complete. Entire speeches, papers, even major reports, can be and are written entirely in 'energese'. Nevertheless, familiarity with these key expressions and their various ambiguities will enable you to make working translations from 'energese' to English, and to understand better the hidden agendas that so often underlie policy pronouncements.

energy: oil; commercial fuels; electricity; commercial energy carriers of all kinds; all energy carriers, commercial or otherwise; ambient energy (see also); ambient energy hardware (solar panels, wind generators and so on); intermediate energy hardware (refineries, power stations and so on); combinations and permutations of the above; almost never 'energy' in its strict physical scientific sense.

ambient energy: energy that is present but unnoticed, usually unmeasured and free of charge, for instance the energy of sunlight, wind, warm bodies and other energy systems warmer than their surroundings.

fuel: 'material for fireplace'; material whose energy content can be mobilized where and when it is desired for use; note, however, that some commentators use 'fuel' to mean only commercial fuel, or to include also electricity, which cannot in practice be stored.

energy hardware: concept conspicuous by its absence in most energese: device or system to intervene in and control the conversion of energy for human use.

power: grid electricity; all electricity, grid or otherwise; used all too frequently as equivalent to energy (see also); occasionally, albeit rarely, used with physical scientific meaning, that is, rate of energy conversion – energy converted per unit time, for example joules per second or watts.

energy carrier: a material or phenomenon that can store energy or transport it from place to place; all fuels plus electricity.

energy production: extraction and processing of fuel; intermediate conversion of fuel into a secondary energy carrier like electricity; controlled conversion of ambient energy, for example generation of hydroelectricity or wind electricity; combinations and permutations of the above; literal meaning is scientifically wrong.

energy producer: see 'energy supplier'.

energy consumption: amount of fuel used, directly or indirectly; amount of fuel plus electricity used; amount of electricity used; amount of commercial energy carriers used; amount of energy carriers, commercial and non-commercial, used; literal meaning is scientifically wrong.

energy supply: fuel available for use; fuel plus electricity available for use; commercial energy carriers available for use; all energy carriers available for use; may also include ambient energy consciously converted for use; does not – repeat not – embrace the energy conversion that takes place without human intervention which makes the earth habitable and constitutes more than 99 per cent of the energy conversion taking place on it.

energy supplier: supplier of commercial fuel or electricity.

energy demand: in the past, recorded purchases of fuel or fuel and electricity; in the future, anticipated purchases of fuel or fuel and electricity; energy carriers converted by final users; can be specified before intermediate conversion, for instance power station, or after, leading to very different results; note that in the future by definition unsatisfied 'demand' cannot exist since 'demand' is identified and quantified only by satisfying it.

energy source: fuel; electricity (a questionable usage at best, and in the case of electricity generated from fuel, scientifically wrong); active ambient energy hardware, for instance solar panel – but not usually the ambient energy itself and almost never the sun, although the sun is the source of almost all the energy converted on earth.

primary energy: virgin fuel; hydroelectricity or nuclear electricity sent out from generating stations; sometimes stated net of energy used in production, sometimes not; sometimes includes non-commercial fuel, sometimes not.

secondary energy: fuel or electricity produced by intermediate converter (refinery, power station or such); sometimes stated net of processing and transmission losses, sometimes not.

delivered energy: amount of energy carrier reaching customer's meter, weighbridge or such; amount of energy carrier for which customer must pay supplier.

useful energy: energy whose conversion identifiably furthers user's objective; delivered energy less energy lost by end-use hardware (up the chimney, in friction and such).

non-renewable energy/non-renewable energy source: fossil fuel; not usually applied to deforestation for firewood, or siltation of hydroelectric installations; sometimes applied to nuclear energy generated by burning uranium in conventional nuclear reactors.

renewable energy/renewable energy source: sunlight and its derivatives, sometimes embracing the relevant hardware, sometimes not; the ambiguity is unfortunate, since most so-called 'renewables' are based on the conversion of ambient energy that is itself free; the costs arise in controlling its conversion for use, for example by wind generators or solar cells.

alternative energy/alternative energy source: see 'energy source'; 'alternative' means excluding coal, oil and natural gas, and may also exclude hydroelectricity and nuclear energy, depending on the commentator; small hydro and advanced nuclear reactors like 'fast breeders' are commonly classed with the 'alternative sources'.

high/low energy: the labels 'high-energy' and 'low-energy' are commonly applied to predictions or scenarios for future energy use, in which the anticipated use of commercial fuels and electricity is 'high' or 'low' compared to some reference level; 'high' and 'low' do not, however, refer to the amount of controlled energy conversion, since 'low-energy' scenarios commonly assume a considerable increase in controlled conversion of ambient energy, and more efficient use of commercial energy carriers.

efficient/efficiency: desirable attribute of energy system in which final 'energy service' is provided by optimum combination of energy hardware, ambient energy and energy carriers; may also, however, be used as all-purpose hooray-words in policy pronouncements, and become roughly equivalent to 'NEW IMPROVED' as used by detergent manufacturers.

energy conservation: 'energy' is invariably conserved in any conversion process, whether or not it is controlled by people; as used in policy pronouncements 'energy conservation' usually means 'energy carrier conservation', that is, using less fuel or electricity, as a result of various measures, negative or positive, short-term or long-term; all too often merely hooray-word, otherwise undefined; best regarded warily.

energy service: what energy-users actually want – comfort, illumination, cooked food, mobility, and so on; providing such 'energy services' with the optimum combination of energy hardware, ambient energy and energy carriers is the key to the 'energy alternative'.

APPENDIX B

ENERGY INFORMATION

Anyone interested in energy today need look no farther than the daily newspapers. Popular magazines like *New Scientist* carry the latest information about energy and environment; and specialist journals like *Energy Policy* (Butterworth) and newsletters like *Energy Economist* and *Power In Europe* (Financial Times Business Information, London) now abound. Governments everywhere publish up-to-date reports on relevant topics, as do parliamentary and congressional committees and international organizations.

Titles on a typical energy bookshelf might include many of the following; they in turn point to many others:

ENERGY STRATEGY: GENERAL

Energy: Global Prospects 1985–2000, McGraw/Hill for Workshop on Alternative Energy Strategies, 1977
World Energy: Looking Ahead to 2020, IPC Science and Technology Press for the World Energy Conference, 1978
Energy In Transition 1985–2010, W. H. Freeman for US National Academy of Sciences, 1980
Energy in a Finite World, Wolf Haefele et al.; International Institute for Applied Systems Analysis, 1981
World Energy Strategies, Amory B. Lovins; Ballinger, 1975
Soft Energy Paths, Amory B. Lovins; Penguin, 1977
Least-Cost Energy: Solving the CO2 Problem, Amory B. Lovins, L. Hunter Lovins, Florentin Krause and Wilfrid Bach; Brick House, 1981
Building a Sustainable Society, Lester Brown; W. W. Norton, 1981
State of the World, Lester Brown et al.; W. W. Norton for Worldwatch Institute, annually from 1984
Worldwatch Papers, Worldwatch Institute, numbered series, many on energy, e.g. 91: *Slowing Global Warming: A Worldwide Strategy*, October 1989
Energy Use: the Human Dimension, edited by Paul C. Stern and Elliot Aronson; W. H. Freeman for the US National Research Council, 1984
Energy for a Sustainable World, Jose Goldemberg, Thomas B. Johansson, Amulya K. N. Reddy and Robert Williams; John Wiley, 1988
Energy for a Sustainable World, Jose Goldemberg, Thomas B. Johansson, Amulya K. N. Reddy and Robert Williams (abridged); World Resources Institute, 1987
Energy for Development, Jose Goldemberg, Thomas B. Johansson, Amulya K. N. Reddy and Robert Williams; World Resources Institute, 1987

ENERGY STRATEGY: US

A Time to Choose: Final Report by the Energy Policy Project, Ballinger, 1974

Energy Future, edited by Robert Stobaugh and Daniel Yergin; Random House for the Energy Project, Harvard Business School, 1979
Energy Strategies: Toward a Solar Future, edited by Henry W. Kendall and Steven J. Nadis; Ballinger, 1980
Our Energy: Regaining Control, Marc H. Ross and Robert Williams; McGraw/Hill, 1981

ENERGY STRATEGY: SWEDEN
Energy in Transition, Mans Loennroth, Thomas B. Johansson and Peter Steen; University of California Press, 1980
Solar versus Nuclear: Choosing Energy Futures, Mans Loennroth, Thomas B. Johannson and Peter Steen; Pergamon, 1980

ENERGY STRATEGY: BRITAIN
Fuel's Paradise, Peter Chapman; Penguin, 1975
A Low Energy Strategy for the United Kingdom, Gerald Leach et al.; Science Reviews for the International Institute for Environment and Development, 1979
Energy-Efficient Futures, David Olivier et al.; Earth Resources Research, 1983

FUELS AND ELECTRICITY
Coal: Bridge to the Future, Ballinger for World Coal Study, 1980
Coal Combustion and Conversion Technology, David Merrick; Macmillan, 1984
Advanced Coal-use Technology, Walter C. Patterson; Financial Times Business Information, 1987
Coal-Use Technology in a Changing Environment, Walter C. Patterson; Financial Times Business Information, 1990
Nuclear Power, Walter C. Patterson; Penguin, 1986
The Realities of Nuclear Power, S. D. Thomas; Cambridge University Press, 1988
Renewable Energy, Bent Sorensen; Academic Press, 1979
Solar Prospects, Michael Flood; Wildwood House for Friends of the Earth, 1983
Renewable Energy, Daniel Deudney and Christopher Flavin; W. W. Norton for Worldwatch Institute, 1983
Electricity: Efficient End-use and New Generation Technologies and their Planning Implications, edited by Thomas B. Johansson, Birgit Bodlund and Robert Williams; Lund University Press, 1989
Solar Hydrogen: Moving Beyond Fossil Fuels, Joan Ogden and Robert Williams; World Resources Institute, 1990

ENVIRONMENT
Acid Rain: Rhetoric and Reality, Chris C. Park; Methuen, 1987
Air Pollution and Acid Rain, Alan Wellburn; Longman, 1988
Climate, History and the Modern World, H. H. Lamb; Methuen, 1982
Turning Up The Heat, Fred Pearce; Paladin, 1989
The Greenhouse Effect, Stewart Boyle and John Ardill; New English Library, 1989
Energy Policy in the Greenhouse, Florentin Krause, Wilfrid Bach and Jon

Koomey; International Project for Sustainable Energy Paths/European Environmental Bureau, 1990

VARIOUS TOPICS
Energy and Food Production, Gerald Leach; IPC Science and Technology Press, 1976
Annual Review of Energy, Annual Reviews, annually since 1975
Handbook of Industrial Energy Analysis, I. Boustead and G. F. Hancock; Ellis Horwood, 1979
The Stove Project Manual, S. D. Joseph, Y. M. Shanahan and W. Stewart; Intermediate Technology, 1985
Household Energy Handbook, Gerald Leach and Marcia Gowen; World Bank, 1987
Solar-Powered Electricity, Bernard McNelis, Anthony Derrick and Michael Starr; Intermediate Technology/UNESCO, 1988
Electricity for Rural People, Gerald Foley; Panos Institute, 1989

APPENDIX C

ORGANIZATIONS

Major participants in energy decisions include international organizations; national governments, their departments and agencies; and major corporations. Many provide information on energy issues, in the form of reports, leaflets and other publications, and advertising. International organizations include the United Nations Environment Programme, the United Nations Economic Commission for Europe and equivalents, UNESCO, the World Meteorological Organization, the OECD International Energy Agency, and the European Economic Community. International funding agencies include the World Bank, the Asian Development Bank, and equivalents.

National governments may have a Department of Energy or equivalent; an Energy Technology Support Unit (Britain) and/or Energy Efficiency Office (Britain), Agence Français pour la Maitrise d'Energie (France), and/or a Secretariat for Futures Studies (Sweden). They may also have national funding agencies like the US Export-Import Bank, the Export Credit Guarantee Department (Britain) and equivalents; and aid agencies like the Canadian International Development Agency, Swedish International Development Agency, Overseas Development Agency (Britain), USAID and equivalents. Governments also support the International Institute for Applied Systems Analysis, Laxenburg, Austria.

International and domestic corporations supply energy technology and fuels. Local monopolies, state-owned or private, supply gas and electricity, under the control of regulatory bodies like public service commissions. Many suppliers participate in the World Energy Conference. Energy professionals belong to professional bodies like the Institute of Energy, the Institute of Electrical Engineers, the Royal Institute of British Architects or the British Association of Energy Economists, or their equivalents in other professions and elsewhere. Note in particular the following:

World Coal Institute, Vicarage House, 58–60 Kensington Church St, London W8 4DB, UK.

Institute of Petroleum, 61 New Cavendish St, London W1M 8AR, UK.

Gas Research Institute, 1331 Pennsylvania Ave NW, Suite 730 North, Washington DC 2004–1703, USA.

Electric Power Research Institute, PO Box 10412, Palo Alto, California 94303, USA.

World Energy Conference, 34 St James's St, London SW1A 1HD, UK.

Many of the above organizations are now moving away from their traditional stance on energy. The leading proponents of the energy alternative, however, are still found mainly in organizations like the following:

Alliance to Save Energy, 1725 K St NW, Suite 914, Washington DC 20006, USA.

American Council for an Energy-Efficient Economy, 1001 Connecticut Ave NW, Suite 535, Washington DC 20036, USA.

Association for the Conservation of Energy, 9 Sherlock Mews, London W1M 3RH, UK.

Beijer Institute, Royal Swedish Academy of Sciences, Box 50005, S-104 05, Stockholm, Sweden.

European Environmental Bureau, rue Vautier/Vautierstraat 29, B1040 Brussels, Belgium.

Friends of the Earth International (and its national affiliates), 26–28 Underwood St, London N1 7JQ, UK.

Greenpeace International (and its national affiliates), Keizersgracht 176, 1016 DW Amsterdam, The Netherlands.

Institute for European Environmental Policy, 3 Endsleigh St, London WC1H 0DD, UK; 55 rue de Varenne, S-75007 Paris, France; Aloysschultestr 6, D-5300 Bonn 1, Federal Republic of Germany.

Intermediate Technology, 103/105 Southampton Row, London WC1B 4HH, UK.

International Institute for Environment and Development, 3 Endsleigh St, London WC1H 0DD, UK; Piso 6, Cuerpo A, Corrientes 2835, 1193 Buenos Aires, Argentina.

International Project for Sustainable Energy Paths, El Cerrito, California 94530, USA.

Lawrence Berkeley Laboratory, Berkeley, California 90/3125 University of California 94720, USA.

Oak Ridge National Laboratory, Oak Ridge, Tennessee 37831, USA.

Panos Institute, 9 White Lion St, London W1 9PD, UK; 1409 King St, Alexandria, Virginia 22314, USA; 31 rue de Reuilly, 75012 Paris, France.

Rocky Mountain Institute, 1739 Snowmass Creek Rd, Old Snowmass, Colorado 81654, USA.

Worldwatch Institute, 1776 Massachusetts Ave NW, Washington DC 20036, USA.

World Resources Institute, 1735 New York Ave NW, Washington DC 20006, USA.

World Wide Fund for Nature International, World Conservation Centre, Avenue du Mont Blanc, CH1196 Gland, Switzerland.

INDEX